Memoirs
of a
Mud-Puddler

by

Michael F Ridd

This book is published by
Grosvenor House Publishing Ltd
28–30 High Street, Guildford, Surrey, GU1 3HY.
www.grosvenorhousepublishing.co.uk

A CIP record for this book
is available from the British Library

ISBN 978-1-905529-48-3

Jacket illustration: Southeast Thailand, November 1968

In memory of my parents,
Donald Frederick Ridd and Stella Joyce Ridd,
who wrote unfailingly when I was abroad, and kept my letters home
– as my mother wrote on the bundle of letters:
"for when he writes the book of his travels".

Preface

What vanity, you'll think, to write about my life as if I were a celebrity, or as if I were rich or particularly successful. I'm none of these, and yet I found I had an inner urge to set it out in print. Not so much a wish to tell others of the life I've led (although this may be of some interest to my friends and family), but more the feeling as I approach the buffers of my life that I wanted to recapture the highlights of the journey.

It was only after I'd written a brief synopsis of my career for The Geological Society that the idea occurred to me. Members are encouraged by the society to do this to assist their eventual obituarist. Without the benefit of anything of that kind I had written several obituaries of departed friends and colleagues over the years, and so I saw the sense of writing an outline of my own geological career and I duly sent it off to the society. But once I'd done that, I found myself returning to it in my computer from time to time and tinkering with it, adding a paragraph or two here and some additional detail there, until there was no escaping the fact that I had embarked on writing my autobiography.

To be exact, I describe in these pages only the first part of my life – admittedly the largest part – which ended with the dawning of the new Millenium. What has happened to me since then I see differently; I see it as my on-going life, and I am less inclined to write about it. I hope Mikiko will forgive me for stopping the story just as she stepped into it; my third and final wife, this book is for you.

Michael F. Ridd
Putney
2006

Chapter One

Driving south in the hot summer of 1976, we broke our journey in Dumfries in the Scottish Borders. After a search through the older part of town, among streets of red-sandstone houses we found what I was looking for. York House was now a BMW car showroom, but just visible above the door were the painted-out words *MacKnight and Ridd Limited, Main Ford Dealers*. "That's where your husband was conceived," I said to Song, pointing to the small upstairs flat.

My father and mother were living there in the autumn of 1935. He was a motor engineer, and by all accounts a very talented one, and had moved to Scotland with my mother to start a garage business. He had obtained financial backing from a canny Scottish businessman and it had gone well at first. But after a while strains developed between my father and his business partner and they fell out. So, at the beginning of 1936 my parents with my older brother John returned to live in Essex, the county in which they had married and spent the earliest years of their married life.

My mother was relieved to be away from the alien ways of Scotland, and my father was, I imagine, sadder but wiser from his first and only brush with entrepreneurial business. Back south again, life was bright and happy for them, my father with a new job at Fords in Dagenham. I was born in a small private nursing home in Gidea Park on 16th June of that year.

Our house was a small suburban semi-detached one, number 51 Fairholme Avenue. At the end of the garden a grassy embankment ran down to the LNER main line, with steam trains rushing from Liverpool

Street to East Anglia and the Essex coast. Apparently even at the age of two or three I had an instinct to roam, for in later years my mother would like to recount to friends how I went missing and was discovered several hours later being carried on a policeman's shoulders near Romford Station. It seems I got there by traversing nearly a mile of railway embankment. While I don't remember the incident, my earliest memories are nevertheless of Gidea Park. I remember my father arriving home from Fords of an evening, and my running along the pavement to greet him as he parked his black Ford Popular by the red pillar-box.

With the outbreak of war my father was assigned by his company to a job which took him the length and breadth of Britain, diagnosing and fixing problems in Ford-powered military vehicles and boats. It meant being away from home during the week and he was concerned for my mother and for us boys being on our own, with the worrying possibility of German bombing raids on the London area.

So it was decided that we should leave Fairholme Avenue and move to Shrewsbury, where my mother's sister lived. I was to form the vanguard of this move, and I remember the journey with my father. Before leaving Paddington he led me to the front of the train and asked the engine driver if I could go up onto the footplate. Among the levers and dials, the hissing steam and the heat, I watched in wonder as the driver and fireman cooked eggs and bacon for their breakfast, on a shovel pushed into the firebox. On the journey north we were served luncheon in the dining-car and watched the fields and towns slip past the window. The seeds of a life-long affection for railways were sown. To this day I enjoy long train journeys, and if I can linger over a meal while watching the scenery roll by, then so much the better.

Auntie Kath lived in a small new development on the northwest edge of town and we celebrated my fourth birthday, the two of us, on a sunny hillside among the dog daisies and the buttercups, eating sandwiches and cake from a red Oxo tin. Uncle Carl worked as a storeman at the Sentinel factory where production had gone over from steam-lorries to army vehicles. But the Sentinel lorries were still a common sight in Shrewsbury in those days – lumbering, snub-nosed trucks with smoke billowing from their cab-top chimneys. It wasn't till later that my uncle and aunt had a child of their own, an adopted girl who became my cousin Valerie, but meanwhile Auntie Kath doted on me.

The new family home my parents had found in Shrewsbury was a spacious modern house at the end of a leafy lane on the edge of town. It hadn't a number and so, out of nostalgia for the Essex street they'd left behind, my parents gave it the name "Fairholme". For John and for me the house, but more so the setting, were wonderful. Across the lane and down a sandy bank was a large pond with plentiful rudd to be caught, while in the overgrown field beside the house was a green and slimy pool with a fallen willow tree across it in whose branches we could climb. Sewage from the houses in the lane collected in a concrete cesspit and we discovered we could prise off the cover to peer down into its bubbling awfulness, and the effluent ran into the pool, accounting for its stench, its blackness, and its rich crust of duckweed. It soon acquired the name "Stink Pond" while the more wholesome one in front of the house was simply "The Pond". After the suburbs of Essex we could not have been happier in this rural wilderness.

A short but powerfully-built man, my father had a pugnacious streak and this came to the fore when standing up for his family or for what he considered to be his rights. "An Englishman's home is his castle" was an adage he took seriously. Ours backed onto the main Oswestry road which was screened by a high hawthorn hedge, and soon after our arrival there the local council decided to cut and lay it. After they had done several of our neighbours', my father got wind of what was going on and forbade the workmen to touch our 30 yard length of hedge. And so, for the years we remained in Shrewsbury, our back hedge stood proudly but incongruously some ten feet higher than the rest. That defiance of officialdom characterised him. On another occasion a squad of the local Home Guard were carrying out an exercise and thought they would climb into our garden and through my father's newly-built chicken run. A fierce row ensued which climaxed in him man-handling the would-be soldiers off his property.

My father was a complex character. He was brought up in a strict Methodist household where intolerance and narrow-mindedness held sway, and throughout his life these hung around his neck like an albatross. The inner conflict between those instilled prejudices on the one hand, and kind and generous instincts on the other, shaped his personality.

But if it seems my father was a severe and joyless man, by no means did he lack a sense of humour. His comic bag of tricks included a number of silly sayings. For example, someone had only to comment what a cold day it was for him to respond "It certainly is; it's cold enough for a walking stick!" Or if he heard of a surprising achievement on someone's part, he would endorse it with "Yes, not bad for a hare-lip!" These and several other sayings have firmly established themselves in the family's language and, notwithstanding the political incorrectness of some of them, I often find myself coming out with them; as I do with some of my father's other amusing tricks – amusing to me, at least. Seek to adjust his tie or remove a crumb from his chin and he would pretend suddenly to snap with an audible growl at your fingers, an alarming experience for anyone unaware that it was a joke, but hugely amusing to us boys. Or if he was handing you a pencil or some other small object, he had the magician's dexterity suddenly to palm the object and extend his forefinger, with the result that an unsuspecting person found himself grasping not the object but a fleshy digit.

His social background was white-collar, but toward the lower end of middle class. Although they had no inherited wealth they felt strongly they were a cut above the working class. They worked for what money they had, and while his parents owned their home in suburban Westcliffe-on-Sea, that was the probably the extent of their assets. A university education was beyond the dreams of most of the lower middle class and my father's career qualifications were gained at night-school. His hard-won education, combined with intelligence and considerable manual skills, meant there were few domestic jobs he could not tackle. Unsure of whether Shrewsbury would remain safe from German air attacks, some of the earliest examples of his handiwork I can remember were the concrete and steel air-raid shelter he built in the garden, along with siren suits made from old overcoats for all the family. He loved us boys and after boisterous games together in the evening he would read us a story in bed and then kiss us goodnight. Toys were virtually unobtainable in the shops in those war years, and so when he was home at the weekends he would work late into the night to make us Christmas and birthday presents.

Notwithstanding his kindness and care for us, he firmly believed that when we stepped too far out of line we should be punished. I received several spankings over the years but the only one for which I can remember my misdeed was one involving my use of bad language.

I was perhaps eight or nine and was playing at the edge of the pond. It was tadpole season and I had gathered a number of the wriggling black things in a tiny part of the pond I had dammed-off with clay. With a stick I'd prodded a hole through the clay dam and I was urging the tadpoles to swim through it into the pond beyond. What I didn't realise was that my Uncle Eric, my mother's brother who also lived in Shrewsbury, was standing on the road nearby observing me as I shouted at the tadpoles: "Get through, you little buggers! Go on, swim through, you buggers!"

Uncle Eric reported this to my father and shortly I was summoned. "Michael, come here at once," he called in a stern voice from the house. "Go up to your room." I went up, quaking with apprehension. Sure enough, a minute later he entered my room with a thin wooden rod, an arrow from an old archery set, while Mum pleaded from downstairs that I should be spared. I was ordered to take down my trousers, bend over, and then there was a whooshing noise and my buttocks seemed to catch fire. I screamed in agony as he explained in words strangled with emotion that I must never use such language again. I'm certain that he hated having to punish me that way. It was necessary for my proper upbringing, he thought, and I have never harboured the least resentment toward him for it. Although I have rarely chastised my own children, I remain convinced that as a last resort a firm slap is not unreasonable.

In spite of these occasional punishments, life was good. It was an idyllic habitat for adventurous boys. Apples, and even peaches, could be scrumped from the gardens behind the high brick wall beyond the pond. To navigate around the pond we made rafts, and once, even a canvas-covered boat. And in the hedgerows mature oaks, elms and ash abounded and we became fearless climbers. The banks around the pond were sandy and easily excavated and soon we had dug a network of tunnels. We would squirm our way through their claustrophobic narrowness like moles, and sometimes we would imprison our enemies in them. From time to time the tunnels would collapse and how none of us got killed in them I don't know. We would simply re-roof them with planks or corrugated iron sheets and carry on.

The pond and the stink-pond were endless sources of pleasure. We would watch in disgust as hordes of frogs writhed copulating in the shallow water, and later we would collect their spawn in jam jars. We would fish with worms for large gold and silver rudd which, we soon discovered, were not good to eat. And we would collect moorhens' eggs which were excellent when fried over a camp fire, in spite of coming from nests on the sewage-filled stink-pond. In winter when the frost was hard enough, (and sometimes when it wasn't), the pond would become our ice rink.

For a boy in his formative years the seven that we lived in Shrewsbury could not have been happier. Countryside with endless scope for exploring and adventure spread northward into the level fields and meres of Cheshire, west to the hills of the Welsh Border and south to the distant beckoning Long Mynd and Stiperstones. There were natural hazards aplenty in the countryside, but on the roads, even on the main Oswestry road behind the house, traffic was light and moved at a safe, leisurely speed. And of course the bombing was happening a long way away. My parents therefore saw no reason to try and restrict our movement and so, with our friends, we were free to roam far and wide.

At the age of six or seven I would be away for half a day at a time with my friend Garth Jones on the small, blue, hand-me-down bicycle that had been my brother's. Garth was the older and he did the pedalling while I sat astride the luggage carrier behind. Through winding lanes to Bicton and Nobold, Meole Brace and Bayston Hill we'd ride, stopping here to catch sticklebacks and there to pick damsons from the hedgerows. Or on foot we would explore the high Severn cliffs at The Roughs where sand martins built their nests, and where a boy was rumoured recently to have drowned in a whirlpool. Or closer to home we'd dare each other to cross the Black Bog, a half acre of dark, fenced-off, swampy woodland, where to step carelessly among the tangled alder trees meant sinking chest-deep in peaty sucking mud.

By being allowed to roam we learned how to survive and, subconsciously, what risks we could take. While cuts and bruises were common, generally the worst mishap that befell us was to get soaked feet. "Mum, I've got a boot-full!" we cried plaintively as we returned home after another of our adventures.

Although it was a Huckleberry Finn existence we were by no means free from having to carry out regular domestic chores. Daily, it fell on John and me to feed the chickens, and I can still recall the smell of the steamy mash of boiled vegetable peelings we would lift from the gas-ring, and mix with 'balancer-meal' and yellow Karswood powder. After every meal it was the childrens' job to wash the dishes, and John and I would jostle to get our part of the job done before the other. But Sunday was the worst day of the week, when we would be forced to put on our best clothes to attend Sunday School. And if that was not bad enough, the household's dirty shoes would have to be polished when we returned home.

Mother was a dear, sweet, woman whose daily round consisted of cooking us nutritious meals, knitting jumpers, bandaging our cuts, and making sure we always left the house in clean clothes. She was never cast down by our routinely returning in them, dirty and torn. It was her nature never to think ill of anyone, least of all her two boys. We must have sorely tried her patience but she never did more than wag her finger at us in admonishment, or if we drove her to her wits' end she would utter "just you wait till your father gets home!" a threat she never carried out.

She was six years younger than my father and was working in Westcliff-on-Sea as a telephone operator when they met. Her father, Henry Granfield, was a civil engineer whose job had taken him all over the country, supervising the building of power-stations. A sturdily self-sufficient man with an intellectual bent, I admired him particularly for his daily year-round swim in the sea off Felixstowe, a regimen he kept up until shortly before he died in his seventies. My mother's broad-minded and tolerant nature probably came from him.

The folks of this green and pleasant segment of Shrewsbury's suburbs sent their children to Pengwern House, a small dame-school a half-mile walk from home, run by an elderly spinster, Miss Toddington. The alternative would have been the council primary school with the rough and frightening children from Frankwell, a crowded area of older housing close to the town centre. Sending us to Miss Toddington's must have involved some sacrifice on the part of my parents, but politeness and speaking properly were even more important than academic attainment, and were matters my father had strong views about.

If the children of Frankwell were rough and working-class, then those at Pengwern House were no angels either. Cherry trees lined the road we took to school, and every summer we would tear down the branches, filling our pockets, satchels and gas-mask cases with the fruit. When bored during classes I liked to illustrate the inside covers of my exercise book with rude sketches of naked backsides and male genitalia, which, when I was eventually found out, earned me a severe rebuke from Miss Toddington. But I shared her revulsion at the far lewder acts that she would discover being committed by older boys beneath the desks at the back of the classroom. One-eyed Miss Gates, however, had no truck with mere rebukes when anyone displeased her during singing or French lessons. She would lash at them with a long cane with a leather-bound knob on its end, and I was frequently among her victims.

Through the summer months we would practice athletics on the field beside the school, culminating in Sports Day with parents present, and the award of prizes. And in the winter we would walk in crocodile down Port Hill, across the suspension bridge (stamping our feet to feel it shake), and through the Quarry to the town baths where I soon learned to swim. After heavy snowfalls in the Welsh mountains the River Severn would flood, and some years it would cause us great excitement by inundating parts of the old town. The suspension bridge would become impassable and the only way into town was through Frankwell, by punt or on temporary boardwalks above the flooded streets.

Shrewsbury, lying only twelve miles from the border with Wales, had a strongly Welsh flavour. Many of its residents spoke with the musical lilt of Wales, and among my friends were boys with characteristically Welsh names, like Richard Richards and William Williams, and of course many with the surname Jones.

My closest friend was John Riley. He had been born with a deformity of the legs which, in his early years, needed periodic operations to make them useable. Eventually he was able to walk, but meanwhile I would push him around in his large, wheeled, push-chair. He was a boy of endless good humour and he would treat his pram as a vehicle in which to have fun, rather than as an impediment. With his encouragement we would fix a rope to his push-chair and it became a chariot, to be bounced over fields and raced at break-neck speed down hills. When inevitably the fun got out of hand and he was pitched out onto the ground he would just climb back in with a laugh. After I moved away from Shrewsbury we stayed in touch by letter for a year or so, but we never saw each other again.

Another boy I met for just an hour or two. It was early morning in the late summer, and I was messing about alone by the ditch that drained the Stink Pond. Then I saw a boy, a stranger, sitting by himself on the bank, half hidden by the long grass. Judging that he was not hostile (for boys are like dogs in this respect) I went up to him. He was of about my own age – perhaps nine or ten – and was dressed, I remember, in a jacket which he wore inside-out. We sat together and talked, and he told me he was travelling, had spent the night sleeping in the open, and that he was hungry. He had collected a handkerchief full of wild mushrooms, and leaving him for a few minutes I went home and returned with matches, a frying pan, and bread and butter. I soon had a fire burning and we talked as we breakfasted together on mushrooms and fried-bread. And then he left. How long he had been on the road I didn't discover, or where he'd come from or was headed. But I felt a deep empathy for him, and the details of that strange encounter stay clear in my memory.

The more I roamed the fields, hedges, woods and brooks of the Shropshire countryside, the greater my interest became in every form of wild animal and plant. I became a passionate bird-egg collector. Before the days of agricultural pesticides and car-borne throngs from the towns and cities, bird life was abundant. My father made us each a wooden cabinet with segmented drawers for our specimens, and over our childhood years they were gradually filled. Most of our collecting was done in the fields and hedgerows near to home, and we developed a strict code of conduct to regulate the number of eggs we could take from a nest, leaving just enough that the parent bird would not abandon it. Exploring with my brother, John, or with a small group of friends, the first to spot a nest would yell out excitedly, "foggy-oggy," which was our childhood code for "first egg," meaning that if fewer than, say, four eggs were present, that person would be the only one allowed to take a specimen. We became expert at piercing the eggs with a hawthorn spine and blowing out the contents, and of course at identifying the bird species. I still treasure the three-volume set, *The Birds of the British Isles and their Eggs* by T.A.Coward, given to me by my parents in 1944, on my eighth birthday.

Occasionally we would discover a nest that required particular ingenuity to reach. One such was the great crested grebe's, a floating mass of vegetation some distance from the shore of Bomere Pool. After much experimentation on the pond in front of the house, John and I fashioned a dinghy from a sheet of canvas stretched over an old tractor inner-tube. This then had to be transported by bicycle the four miles to Bomere, launched under cover of darkness to avoid the landowner, and paddled out to the nest. The expedition was a complete success and resulted in a single white egg the size of a bantam's which we shared and which became the centrepiece of our collections.

It was not just the bird-nesting that we found such fun; we enjoyed the true collector's excitement of building our collections in any way we could. In Cheshire, we learned, there was a commercial dealer called E.G.Gowland, who stocked every kind of rare egg from red-throated divers to golden eagles. We would pore over his catologue for hours before sending off our postal-order, and then count the days till the postman arrived with the wooden box of cotton-wool-packed treasures.

Nowadays egg-collecting is rightly unacceptable, and the law protects Britain's diminished bird life from collectors who are able to scour the countryside. But I have no regrets over my own collecting, carried out in a time when the countryside was still rich and bountiful.

Besides an interest in living things, both animal and plant, I developed a fascination for maps. My father had a number of Edwardian motoring maps from his youth, some of them in the form of route maps. In my imagination I would travel from Edinburgh to London down the Great North Road, or thread my way through the Chilterns and the Cotswolds to Gloucester on the road we now call the A40. But the maps I pored over at greatest length were a Bartholomew's map of Scotland which was dog-eared from my father's wartime travels, and a one-inch-to-the-mile Ordnance Survey map of the Long Mynd which I bought in Shrewsbury for about a shilling. I day-dreamed of visiting the Grampians and Rannoch Moor and, closer to home, of walking The Stiperstones and Wenlock Edge, and they seemed as exciting as if they were in China or deepest Africa. That love of maps has never left me, and perhaps it played a part in channelling me toward becoming a geologist, for whom a map is as indispensable as a tape-measure is to a tailor.

I can't recall riding in a car at any time during the War Years and my father's Ford spent the period on wooden blocks in the garage. We would walk or cycle, or to go into town we would take the bus. And horse-drawn carts were not that uncommon. Somehow I fell in with the man who used to collect kitchen waste from the army barracks in

Frankwell to use as pig-swill, and it was fun to sit beside him on the driver's bench, sometimes taking the horse's reins as we clattered through the town with our drums of sour-smelling swill.

I remember Shrewsbury in the 'forties as a quaint old-fashioned town, nearly encircled by the Severn, with an abundance of ancient half-timbered buildings in alleys and narrow streets with names such as Wyle Cop, Mardol, Shoplatch and Dogpole. Sadly it suffered at the hands of planners in the 'fifties and 'sixties when so many British towns and cities had their hearts torn out and replaced with ugly and frequently jerry-built structures – a national madness that swept the country in the name of the new god, modernity. One of Shrewsbury's greatest losses from that period was the fine Victorian market hall where my mother bought her vegetables and eggs and made friends among the local farmers and their wives. In its place now stands a characterless four-storey office block.

The war years were hard for housewives like my mother, with food rationing only adding to the difficulty of bringing up a family. The one domestic electrical appliance she had to lighten her workload was a cumbersome Hoover. Because our house was large, with two spare bedrooms, relatives from the south would come and stay for weeks or months, adding to my uncomplaining mother's harassment, particularly when quarrelsome cousins were among the guests. But for me the war meant little. I was aware, of course, that the British were fighting the Germans and the Japanese, because I was told to keep quiet and listen, as the successes and set-backs were read out on the six o'clock radio news – the voices of newsreaders John Snagg and Alvar Liddell became as familiar as members of the family. The war also meant that when, on a Tuesday, I bought my tuppenny Dandy comic paper I had to remember to take my "points" if I wanted to buy a bar of chocolate or a quarter of toffees.

I was seven when my sister Judith was born, on 13th November 1943, and I remember diverting on my way home from school so that I could wave to Mum, standing at the upstairs room of the nursing home on Copthorne Road. I well remember my parents' cries of alarm a few months later when I was walking Judith in her pram. I had allowed it to run off the kerb and overturn, rolling Judith into the road, but thankfully without mishap – perhaps the exploits with John Riley and his push-chair were not the best training for my new baby-walking role.

It was around then that soldiers appeared on our beloved fields beyond the pond, and the long, low, barrack huts and workshops of an army camp started to spring up. Soon we would hear the bugle calls of reveille at meal times, and the sound of troops on parade, and heavy army vehicles would churn the fields to mud. But rather than reducing our scope for exploration, it turned out the camp added to it. Through the barbed wire fence there were rich pickings to be had: timber and corrugated-iron sheets for our den-building, fuel drums for making rafts, roofing-felt to be melted over an open fire to collect its tar, and even a Canadian-style canoe which we commandeered for several days until the Army discovered it on the pond and took it away.

For some weeks I had been an observer as John worked on a canvas-covered canoe of his own that he was building, a craft of Heath-Robinson design. With any holes in the ancient canvas sealed with molten tar, it was now complete and ready for its maiden voyage on the pond. But meanwhile the Army had taken to using the pond for washing their tracked Bren-gun carriers, and inevitably the water had become polluted. Large numbers of dead fish appeared belly-up in the reeds around its margin. Not put off by this, or by the Army's presence a few hundred yards away, we dragged John's new canoe from the garden and eased it into the water.

Miraculously the flimsy craft remained afloat and I watched enviously as John lowered himself on board and gingerly paddled it into deeper water. Everything was going well, but I remember being distracted by a particularly large dead fish (unidentifiable, of course, in its pale bloated condition) lying in the shallows. I found a stick to drag it to the edge and hauled it onto the bank. John would love to see this, I thought, and I called to him to take a look. Out on the pond, he was concentrating on other matters, and I got no response. And so I picked up the slimy fish and hurled it toward him, shouting "How about this one, John, it's a real whopper!" If he hadn't ducked so sharply he would have been alright. In avoiding my flying fish, his canoe capsized and tipped him into the water. I suppose if I'd been a more devoted brother I would have stayed around to offer help. But prudence and thoughts of self-preservation overcame me, and I retired. From a safe distance I watched as he swam to the fish-smelling shore, dragged himself out, and trudged dripping and cursing home.

I was nearly nine when the war in Europe ended. The family were holidaying with my father's parents in their large, dismal, Victorian house in Westcliffe-on-Sea; (why we were not at school in Shrewsbury, I can't remember.) It's dark and fusty interior somehow matched the stiffness and strict Methodism of Grandpa and Grandma Ridd. The neatly-trimmed suburban garden offered little scope for the pleasures we were used to in the Shropshire countryside, although trips to the Kursaal fairground, or by electric train to the end of the pier at nearby Southend, provided some relief from the enforced gentility. On the day that victory in Europe was declared, 8th May 1945, we were staying briefly with my father's sister, Auntie Mabel, and her family in Ilford. Leaving Judy in their care, my parents, John and I took the train to London to share in the celebrations. I can remember my father carrying me on his shoulders down the Mall, and with thousands of others packed tightly in front of Buckingham Palace, we clapped and cheered the Royal Family as they stood and waved from the balcony.

Back in Shrewsbury, the end of the war meant that the car could be lifted off its blocks and brought into use again. My father was now commuting weekly to Dagenham, and at weekends he would take us on outings, driving perhaps to Dudley Zoo, and on one occasion to the music-hall at far-off Wolverhampton (or was it Newcastle-under-Lyme?), where George Formby with his ukelele was top of the bill. There were trips as far afield as Lake Bala and Betws-y-Coed in North Wales, and nearer to home the ancient whaleback massif of the Long Mynd became a frequent day out.

The winter of early 1947 was unusually harsh and snow lay on the ground for weeks. The fine toboggan my father had made was towed behind the car to Cardingmill Valley where we careered down the flanks of the Long Mynd. And I remember a firewood-gathering trip to the nearby Black Bog, dragging the toboggan – in what my mind's eye sees now as if it were a Breughel painting – and ending with the ice collapsing beneath me, and the trudge home in drenched clothes which hardened in the freezing air.

Chapter Two

There was no longer any reason to remain living in Shrewsbury, and my parents set about finding a home for us back in Essex. They chose a pretty semi-detached pre-war house of unusual oak-beam construction in Harold Wood, just a couple of miles from Gidea Park where we had lived before the War. It was a pleasant leafy village, and the Ridgeway, running off Gubbins Lane, was an un-made gravel road flanked by wheat fields. We arrived there on a fine sunny day in early summer, the summer of 1947. It was holiday time, and John and I spent the days exploring the rolling countryside which spread from our very front gate. Geoeffrey Ball, the boy next door, was my own age, and we soon became the best of friends.

According to my father, the Ball family were socially acceptable, as was Mrs Husk a few doors up the road (whom my mother often referred to puzzlingly as being "a clean little body"), whereas the Triggs at No 5 were, in his view, rather low class, and we were discouraged from playing with their boy. Similarly he looked down on several of our other neighbours, and I believe this was largely to do with their accents. We had left behind the Welsh lilt of Shrewsbury, and now found ourselves in a neighbourhood where "estuary English," if not downright cockney, was the norm. Remarkably, considering my father's lower middle-class background and his upbringing in the East End and Westcliffe-on-Sea, he spoke with received pronunciation. In bringing up his own children, one thing he made sure of was that they should speak the same way. Hardly a day went by that he didn't correct us for our glottal stops or our dropped aitches.

John had been at the Priory School in Shrewsbury, and I too had just passed the 11-plus exam, and was due to begin at grammar school when the holidays were over. The school selected by my parents was the Royal Liberty School in Gidea Park. We started there in September, dressed in our blue caps and tweed jackets, Dad considering that the latter would be longer-lasting and better value than the prescribed blue blazers. John was put into the Third Form and I went into Form 1C, the stream for the less academically inclined. No doubt Miss Toddington had sent the Headmaster, Mr.Reg Newth, full details of my school career at Pengwern House, including her sobriquet for me, the "mud-puddler". It was not so much a nick-name as a generic term she used for children like me whose preferred habitat was the fields and ponds and woods that we loved to mess about in. I suppose I was rather flattered to be a mud-puddler, thinking of it as a badge of honour. She wrote a charming letter to my parents in which she said:

"I shall miss Michael very much. He is an unusual kind of boy and I have always liked him and especially so since he has been in my form. He is showing more and more intelligence in his work and I feel sure he will do well in his new school."

I soon settled in among the thirty or so boys in Form 1C, not distinguishing myself at any of the subjects except woodwork and biology, and disliking organised football and cricket. I joined the school's Boy Scout troop, the Twelfth Romford, and I remember enthusiastically pulling the trek-cart, heavily laden with tents and other gear, the few miles to Havering on my first camping expedition. And later there was a camping trip to County Meath in Ireland. That was my first journey outside Britain, and so was a big adventure. Apart from the excitement of the sea voyage from Liverpool, the things I can recall most vividly are the sweets we could buy without the need for ration-cards, and being intrigued by the bare-footed Irish children.

We used to cycle to school each day, a fifteen-minute ride – unless we were lucky enough to get into the slipstream of a passing bus, which meant the trip took only ten. We soon discovered that Fairholme Avenue was just a short bike ride beyond Gidea Park railway station, and so John and I rode their one day after school, to look at the house where we had lived before the War. We knew that our friends, the children of the West family at number 55, had all been killed by a German bomb but we weren't prepared for what we saw. The line of houses backing onto the railway embankment had a gap where number 51 and its neighbours had stood. The only sign that our house had ever been there was the front door-step, just visible among the willow-herb and the long grass. As an eleven-year-old, this made a powerful impression on me, one I can still recall clearly.

In 1948 I moved up into Form 2S. This was the Spanish stream, and was for boys who would struggle with the more difficult languages, German or Latin. Teachers were in short supply and presumably schools had to take whoever was available, but looking back it is hard to see how some of them managed to hold onto their jobs. The history teacher, when displeased, would stomp down the aisle, lift the offending boy (myself, as often as not) out of his desk by the scruff of his neck, and then hit him about the head with both hands. And the PT teacher whacked boys' buttocks with a plimsoll so often that I wonder now if he was in fact a sadist. But sprinkled among this mixed bag of teachers were a few who were good at their job and psychologically normal and so who earned our respect and admiration.

I had dropped out of the Boy Scouts by this time, disliking the church parades which had become a frequent feature. But I remained passionate about camping. One summer with a couple of friends we set off by bike on our own camping trip to Devon, pulling our gear in a trailor which my father had made. Curiously, the only thing I remember of that expedition is the journey home. After striking camp outside Frome on the last day, we set off eastwards. But by the time we reached Devizes my two friends had got the smell of home in their nostrils, and as I pedalled on across the chalk hills of the Marlborough Downs I watched their backs disappearing into the distance. It remained for me to haul the trailor the entire 140 mile journey back, through Newbury, Reading, Maidenhead, Slough and London to Harold Wood, eventually easing myself out of the saddle at about midnight. It's strange that I cannot now remember where we went on that trip, and I wonder if it is because of amnesia brought on by the gruelling journey home.

Although by my teens I was no longer an egg-collector, I remained very interested in every form of wildlife, and particularly in birds and bird-watching. There was a group of Sixth-Form boys who were equally keen and much more expert than I, and they allowed me, still only a Fifth-Former, to join them on their bird-watching expeditions. On winter weekends we would leave home before dawn on our bikes, meet on the Southend Arterial road, and then head off to the Essex coast. Armed with field-glasses from an aged relative, sandwiches and flasks of hot soup, we would tramp along the grassy seawalls, making notes on the flocks of migratory waders and wildfowl that fed on the mudflats and salt-marshes. These were generally one-day trips and it would be after dark before we got home, tired, cold and muddy. But sometimes we would stay overnight in an old hut on the Bradwell seawall, sleeping among the spiders on the dirty, straw-strewn floor.

The main figures in this ornithological gang were Derek Gobbett and

David Bridgewater, and a third was Roger Hurding. One incident on an expedition I remember well – we awoke in our hut on the seawall to find Roger was behaving strangely, and his unnatural appetite for golden syrup at breakfast puzzled us. Later in the day he collapsed from utter exhaustion and, after contacting his parents to get him driven home by car, he remained off school for some time. Roger went on to become a doctor, and many years later he wrote his autobiography [1] in which he described that occasion as the day he discovered he was a diabetic, a disorder which sadly blinded him eventually.

Expeditions further afield included a cycling trip to the north Norfolk coast to see the rare avocet, and a particularly cold week one winter on the Isle of Wight. That trip to the Channel was among my earliest brushes with geology. Gobbett and Bridgewater were studying A-level geology in the Sixth Form, and I became totally enthralled as I helped them dig for ammonites, fossilized bits of bone and petrified wood in the sea cliffs along the island's south coast. The trip was memorable also for its last day, an arduous ride home from the Youth Hostel at Holmbury St Mary in the Surrey Hills, through deep and freshly-fallen snow. We generally stayed in Youth Hostels on these bird-watching and geological expeditions, and soon I became a keen member, assembling a large collection of hostel stamps in my YHA membership card.

Collecting butterflies and moths became another hobby. Geoffrey Ball and I would head for the woods of a summer's evening, with our bottle of beer-and-syrup bait which we would smear on the tree trunks. Later we would do the rounds of the trees by torchlight, deciding which of the drunken moths were to be tapped into the killing bottle. It was possible then to buy ether and chloroform at any chemist, and we were not averse to experimenting on ourselves with these anaesthetics in the shed at the bottom of my parents' garden. For the specialist equipment such as setting-boards and butterfly nets, the only place to go was Watkins and Doncaster, entomological suppliers who had a shop behind a large butterfly sign off The Strand. Alas, the shop has long since disappeared, as has another mecca for boys, the Bassett Lowke showroom on Holborn, where complete kits of working-model steam trains could be bought by those with more pocket-money than I had.

If these sound like innocent and harmless boyhood pastimes, our exploits with gunpowder were more hair-raising. We were fascinated by explosives. Sulphur and salt-petre could be bought over the counter, and the charcoal we would make ourselves. After grinding the three ingredients in a mortar and pestle we would experiment with different mixes, ramming a quantity of the powder down the barrel of our home-made

[1] "As Trees Walking", published in 1982 by Paternoster Press.

cannon. This device was made from a length of three-quarter inch steel water pipe, one end sealed with a wooden plug and secured with nail rivets. It was clamped with screws onto a wooden block and a fine touch-hole was drilled into the chamber. After the charge of gunpowder had been rammed home, a wad of newspaper followed and then the projectile, which may have been a bolt, or if we could find one, a steel ball-bearing. A small cone of gunpowder would be poured onto the touch-hole and the cannon was then ready for firing. The test of its potency was the number of sheets of corrugated iron the bullet would pass through; generally it was several. But if it was obviously dangerous to be in front of it when it was fired, it was only marginally safer to be the boy firing it. At times the barrel of the cannon would tear itself off the wooden block and fly backwards between his legs, and on occasions an unusually rich mix would split open the barrel. But we survived, and the only injury I can remember was sustained by John, who burnt off his eyebrows one day while drying a batch of gunpowder on a tin lid held over our endlessly-useful Primus stove.

Another less-than-innocent activity I carried out with a friend (who had better remain nameless) involved the decaying Tudor manor house we had discovered while exploring the countryside around the picturesque hill-top village of South Weald, a half-hour by bike from home. Abandoned and neglected, the house had once been the seat of Sir Anthony Browne, a Sixteenth Century lawyer and Member of Parliament. With its ice-house and underground passages, its cellars and bat-inhabited attics, it offered endless scope for a couple of inquisitive teenagers. While clambering around its stone parapets one day we saw that there were large quantities of lead sheeting and it did not take us long to realise the commercial potential of this find. Raids had to be made under the cover of darkness, of course, and the amount of lead we could carry in a haver-sack hung from a bike's handlebar was not great. Back home we would wait until the coast was clear and then melt the lead in syrup tins on the gas stove. A fishing tackle shop in Upminster bought the resulting lead ingots for ten shillings each and we had made two or three pounds profit before our respective fathers found us out and punished us severely. The house was eventually demolished and I'm embarrassed at the thought that I may have contributed toward its demise.

In the grounds of the Tudor manor house was a large number of ancient oak trees, hundreds of years old and quite possibly planted around the time the house was built. Some of them were completely hol-low and you could climb into their spidery interiors, and high in the crowns of these trees were holes where jackdaws nested. It was my father who first suggested acquiring a jackdaw fledgling and hand-rearing it, and a visit to Sough Weald in April or May became an annual occurrence. We were experts at climbing trees, and so getting to the nests was not difficult. The bird itself had to be of just the right maturity, ideally when it was fully fledged and almost ready to fly from the nest. We would cycle home with Jackie (for that was always its name) tucked inside our jacket, and then commence a regime of feeding it on remnants of our own meals which we would first pre-masticate and then dribble into the jackdaw's beak. A jackdaw soon becomes tame and we would walk around the neighbourhood with it perched on a shoulder like Long John Silver's parrot. But the same tameness which made them ideal pets would also be their undoing. They became a nuisance to our neighbours and would scare old ladies and alarm mothers when they landed on their babies' prams. It was then time to part. My father would drive us and Jackie to a suitably distant wood. Jackie would be encouraged to leave our shoulder and then we would drive away fast, leaving the bird to its own devices.

Throughout my first five years at the grammar school I was an indifferent scholar and had dropped from my curriculum any courses which I found boring or difficult. When the time came to sit my O-Level exams, in the summer of 1952, I took just six subjects: French, English, Chemistry, Biology, Geography and Mathematics. I was awarded a pass in all six but was successful in Maths only because my father insisted on private tuition after I had failed that subject in the February mock exam. As my tutor explained the fundamentals of arithmetic and algebra in the weeks leading up to the exam, the scales fell from my eyes; by June when the time for the exam arrived, they no longer held for me their former terror. If there are decisive points in a person's life, that was one of mine; my career would have had to follow a very different course had it not been for my father's farsightedness.

With six O-Level passes behind me my father considered I was ready to leave school and get a job. That he realised the importance of education could not be doubted, but he had no experience of higher education; Ridds had always left school at sixteen, entered a trade or profession, and gained their qualifications through an apprenticeship or by becoming articled. This route had been followed by my brother John, who left school after the fifth-form and joined a firm of quantity surveyors in London, studying in his own time. And so I had an uphill battle to persuade my father that I wasn't simply being perverse, and it caused a good many stand-up rows between us. In the end it was a meeting he had with the headmaster which changed his mind. Newth was a small, dome-headed man with slicked-down hair and a way of talking out of the side of his mouth. No one took liberties with this man, who ruled the school, if not with a rod of iron then with a rod of bamboo. The next day he called me into his study. To my immense pleasure he explained that my father had relented. Then he dismissed me, but as I turned to go, he asked, rhetorically: "Ridd, do you know what the letters T A C T mean?" As the headmaster spelled out the letters my mind went blank; I hadn't a clue what he was talking about, and I looked back foolishly at him. But as I walked off down the corridor the penny dropped – fifty or so years later I am still trying to control that confrontational instinct, a tendency to tactlessness which has caused me so much unnecessary friction.

The move up from fifth to sixth-form required as big an adjustment as it had done to enter grammar school in the first place. The science subjects I chose to study at A-level were uncompromisingly full of technical terms, formulae and equations, which needed to be memorized to become part of our everyday language. Suddenly I was in an adult world where no one but myself would decide how hard I should work. It seemed obvious that I should take zoology and botany as they were extensions of the O-level Biology I enjoyed, and they were taught by an enthusiastic young teacher whom I liked, called Monty Faithfull – the one teacher with whom I have stayed in touch till this day. My bird-watching friends, Gobbett and Bridgewater, had introduced me to rocks and fossils and I found myself becoming more and more attracted to the subject of geology. Perhaps because it involved maps, travel, the outdoors and natural history, it seemed the ideal third A-level subject. And my fourth subject was chemistry, a subject which I enjoyed at O-level but for which I found I had little flair at A-level.

My battles to become a sixth-former were paying off. The zoology and botany were absorbing, but it was the geology that engrossed me most. The geology teacher, Mr Reekie, or Dan as he was nicknamed, had started out as a mining geologist in Rhodesia. But he developed a seious sinus disorder and after major facial surgery he was forced to leave that career and return to Britain. It was late in the war and secondary-school teachers were scarce, and before long Reekie found himself recruited to teach biology at the Royal Liberty School. It is a credit to Newth that he soon discovered Reekie's wider abilities and asked him to begin a sixth-form geology course. At that time there were very few schools teaching the subject at A-level, but it soon attracted a number of boys. I remember as an eleven-year-old finding Dan Reekie a frightening person; his operations had left deep clefts above one eye and his voice was a deep nasal drone delivered with a Fifeshire accent. But he was a kind and gentle person and as a teacher of his beloved subject, geology, he was inspirational. Nearly thirty years after leaving school, in

1982, I received news of his death at the age of 82. Boys he introduced to geology had fanned out across the world, and many of them now held senior positions in industry, government service and academia. I was moved to write an obituary for the man who had played such a pivotal role in my own life and it appeared the following year in the Proceedings of the Geologists' Association.

My growing geological knowledge allowed me to make sense of the British countryside and scenery that delighted me so on my cycling expeditions. I could now see how the hills of ancient slate in North Wales relate to the younger rocks of the coalfields in the south, and how the belts of limestone and clay which outcrop across England are just the exposed edges of a series of tilted layers. Essex may not be the ideal county for a young geologist to grow up, but by cycling south to the cement factories along the Thames we could collect fossil sea urchins from the chalk quarries, and there were the Walton Crag shell beds to explore along the county's northeast coast. More prosaically, I would knock flakes off kerbstones in Harold Wood with my new geological hammer, and then grind them to wafer thinness with carborundum powder at the kitchen sink, before mounting them on glass slides to study their mineralogy using the school microscope.

A year after entering the sixth-form the Queen's coronation took place, in June 1953, and I found myself standing in The Mall watching the procession through a cardboard periscope. My mother's friend, Mrs Jones, who worked in the Harold Wood Co-operative grocery store, was a war widow and she had received two tickets entitling her to line the procession route. Her pretty daughter Dawn wasn't able to attend, and so Mrs Jones wondered if I might like to take her place. The rainy weather didn't dampen my patriotic enthusiasm, and I enjoyed the day out, taking pictures of the Queen in her golden coach. But more important, I now had an excuse to visit the Jones' house and become better acquainted with Dawn. She was a sixth-former at the Romford Girls' High School, and had a sweet smile and a generous nature. Before long she was letting me walk her home from the bus-stop after school, and as the autumn evenings darkened, buxom Dawn gave me my first lessons in the rudiments of cuddling and kissing.

In the Summer of 1953 Geoffrey Ball and I decided on a camping trip to Scotland. We arranged a lift to Glasgow with a car-delivery firm known to my father, and from there we hiked along the shore of Loch Lomond. It was my first trip to Scotland and I loved its untamed mountain scenery. Beneath rucsacks heavy with the novice hiker's excess of clothing and equipment, we made camp in woods on the hillside above the picturesque lochside village of Luss. Little did I imagine that 25 years later I would be living nearby, seeing from my kitchen window the same Ben Lomond as I saw then through the flap of the tent. It rained incessantly and the midges swarmed over us inside my old-fashioned tent; we hadn't the money to buy sufficient food, and so supplemented our diet of biscuits and baked-beans with boiled nettles. Geoff became more and more dejected and homesick. The midges forced us to remain inside our sleeping bags, which got wet from the rain dripping from the tent's porous roof, and after several days of this discomfort we agreed we would return south by train. Geoff went straight back home to Harold Wood, but I chose to salvage something of our trip by changing trains at Crewe and arriving the next day in Shrewsbury to stay for a while with Auntie Kath. My parents sent up my bicycle by train and I enjoyed a few days cycling again around my old Shropshire haunts. Instead of cycling back to Harold Wood direct, I decided on a tour of Wales. Brief entries in my 1953 diary show I stayed at youth hostels at Van, and then on to the southwest coast of Wales at Poppit Sands and Pentre Cwrt, before turning east through Tyn-y-Coed, and my last night of hostelling on that trip, at Mitcheldean in the Forest of Dean. Home comforts were by now beckoning and I cycled the whole way home to Harold Wood the next day, coincidentally another 140 mile slog. Not bad, I thought, for someone with the skinniest legs in the Sixth Form.

Highlights of the two-year A-Level course were the Easter field trips. In 1953 we travelled to the village of Austwick in the Craven district of Yorkshire, singing off-colour songs in the back of the coach as we took in the unfamiliar limestone scenery. For the next fortnight Dan Reekie led us over the clints and grikes of Carboniferous Limestone pavements, searching for fossil corals and brachiopods; to the summit of Ingleborough in the snow; and fossicking for copper and lead minerals in the spoil heaps of the defunct Pennine mines. A year later it was the Dorset coast where the Jurassic and Cretaceous rocks are even more fossiliferous than the Carboniferous, and are faulted and folded in dramatic style. Again Dan was a great leader, unflagging and able to fill even the dimmest pupil with enthusiasm. By now there was no doubt in my mind that I wanted to become a geologist.

Before the Sixth Form I paid for my travels around Britain out of my five shillings per week pocket money and the postal orders I received from relatives at Christmas and on birthdays. But as a sixteen-year old my travel ambitions were expanding and I needed to get a job to earn some money. A contact of my Uncle Frank in Ilford owned an ironmongery business and it was agreed that on Saturdays I would work in his shop at Gants Hill. It meant cycling the seven miles each way from home, and I was to receive ten shillings for the 9.00 a.m to 6.00 p.m working day. The proprietor's name was Charles Hughes, a tough Welshman who felt he was doing you a favour if he let you off for half an hour in the middle of the day to cycle to the park for your sandwich lunch. The first job on arrival was to wind out the canvas awning and arrange the dustbins, wheelbarrows, ladders, buckets, rolls of wire-netting and other large items of ironmongery, outside on the forecourt. Serving customers took up most of the day but in slack periods there were deliveries to make on the firm's trade-bike, often with tins of paint, a bucket, rolls of wallpaper or a set of saucepans piled on the platform in front of the handlebars. There was tea to make, and shelves of goods to dust, paraffin to measure into customers' gallon cans, and sweeping the floor no matter that it had been swept only ten minutes previously. It was essential, Mr Hughes felt, to appear busy at all times.

One of the other boys doing the same A-Level science subjects as I, was Ian Rolfe. I had known him in the lower school as a tall, lanky and bespectacled boy, with no interest in cycling or camping that I was aware of, and a bit of a swot. But our A-level subjects brought us together and we became the best of friends. His more serious nature began to rub off on me, and I joined the school choir and had a brief foray into school dramatics, (only to give up after losing my toga and muffing my thirteen lines as Cinna the conspirator in Julius Caesar). I was able to get Ian a job at the other branch of Charles Hughes' ironmongers shop, also at Gants Hill, and so we would cycle there together of a Saturday morning, and occasionally were given a lift home by his father in his glazier's van.

Ian's and my interests complemented each other and I introduced him to the lakes and woods around South Weald – the lure of the decaying Tudor manor house had now waned. One of the big attractions there was the herd of deer which roamed wild through that area. Generally we would leave our bikes under a hedge and head into the woods, ensuring that we stayed clear of any farmers or gamekeepers; but on one occasion we decided we would approach the village squire for permission to study a certain kind of wild orchid, the twayblade, we had seen on an illicit visit growing in his woods. The Colonel had a large house on the hill leading up to the village, and we rehearsed our lines before approaching and knocking at his door. He would be very pleased to allow us to go into his woods, he said, and what's more, he would accompany us. He was a charming old gentleman who, when not up in the City – where, bowler-hatted, he went each weekday – he would stroll his forest rides with a long bill-hook, keeping the undergrowth in check. As the three of us approached the grove where we knew the orchids grew, Ian happened to see something glinting in the long grass.

He bent and, to our astonishment, picked up a gold watch. The Colonel was overjoyed, and explained he had lost it a year or so previously. If he had been friendly toward us before this discovery, now it seemed there was nothing he would not do for us. "In future," he said, "you must come here and enjoy the woods just whenever you wish. But you'll need to know the password," he added mysteriously. Perhaps a half century later I can reveal that it was "Abercrombie" who, we learned that day, was the architect of London's Green Belt policy, something which was understandably close to the heart of an Essex landowner.

That summer I took advantage of this new friendship with the Colonel, and he allowed me to join the team hay-making on his farm. Those June evenings were sweltering hot and the work was back-breaking. Armed with a pitchfork, I was stationed at the foot of the hay elevator. No sooner had I heaved one tractor's load onto the elevator than another arrived, and this went on all evening as the haystacks grew. Finally, as darkness descended and we were ready to drop from tiredness, I remember the Colonel arriving on the scene bearing a large keg of beer. "Well, Michael" he said, "as you've been doing a man's job, you can have a man's drink!" In the Ridd household the only drink allowed was a bottle of sherry – for toasting the Queen on Christmas Day and for occasional visitors – and so beer was a novelty. Dog-tired and dry-mouthed, I had never tasted anything quite so delicious.

It was typical of our wise headmaster, Newth, that he had established a tradition of putting boys forward to take part in the annual expeditions of the British Schools Exploring Society. Founded in 1932 by one of Scott's Antarctic compatriots, Surgeon Commander Murray Levick, the society took about sixty boys each summer to such places as Iceland, Lapland or Newfoundland. Throughout my time at the Royal Liberty I was aware of senior boys going on BSES expeditions, and camping and trekking in such remote and romantic-sounding places was something I was desperate to do. Early in my last year at the school I applied to take part in the society's 1954 expedition, which was to be to Northern Quebec. I was interviewed at their office in Northumberland Avenue, and when I received the letter saying I had been selected I could hardly contain my excitement. It would cost me about £140. I worked at the ironmonger's every day that I could, including the school holidays; Essex County Council contributed; and my father probably made up the last few pounds I needed to reach my goal. Three other boys from our school were chosen to join the expedition, including Don Phipps, a good friend. (Don was later to make his career in Canada as a mining geologist, and after a gap of over fifty years we resumed our friendship, staying in each others' homes in Scotland and Sudbury, Ontario).

I sat my A-level exams in June, and a month later with my school career behind me I joined the Northern Quebec Expedition. Heathrow's passenger facilities were rudimentary and we were seen off from London's Victoria Air Terminal by the founder and by Sir John Hunt, who the previous year had led the successful Everest expedition. As Sir John walked along the line of young explorers, stopping to make an encouraging comment to one boy or another, he looked me up and down and said with a grin "You should have no difficulty carrying a heavy rucksack with those broad shoulders of yours!" It was true. At 18 years of age I was not very tall but I had inherited my father's shoulders. I continued to grow for the next couple of years, by which time I had reached six feet. But my proportions were – and have remained – unusual, with long thin legs, a short body and broad shoulders. Socially, I have always felt at ease when on my feet, but when seated at a table I have often felt at a disadvantage and in need of a booster cushion to sit on!

It was my first journey by plane and I can recall the roar as the Lockheed Constellation ran up its engines before take-off, shaking and straining like a leashed bull terrier, flames belching from its red-glowing exhausts. After a re-fuelling stop in Iceland, we landed at Dorval Airport in Montreal where we were entertained by the city fathers before entraining for the over-night journey northeast into the wilderness. If the flight from London were not excitement enough for an eighteen-year-old who had never been in a plane before, the flight from the railhead to basecamp on Lake Wakonichi was an even bigger adventure. We were shuttled a few at a time by single-engined float-plane and I was among the lucky ones and found myself up front, seated next to the pilot.

For six weeks we lived under canvas among the boundless forests and lakes of the Canadian Shield. We trekked great distances and survived on a diet based on 'hardtack' ship's biscuits and stuff called pemmican, a gritty meat derivative which was boiled over a log fire to make a hoosh. We were constantly hungry, in spite of the glades of blueberries we came across and the occasional trout we were able to catch. The organisers had advised us to bring mosquito nets to cover our heads, but labouring over rough ground strewn with fallen timber, and under heavy rucsacks, we sweated profusely and soon dispensed with this cumbersome headgear, with the result that mosquitoes and blackflies attacked us in vast swarms. The basecamp loo – a couple of felled trees over a deep pit – was a favourite haunt of these stinging critters where they could choose any of a half-dozen bare bums lined up every morning.

The society has a good safety record, but there has been one fatal accident in its seventy-four year history, and it occurred on our expedition. One of our members was crossing a lake when a storm blew up and swamped his small boat. For several days we searched the lake shore, but in vain. Then some local Indians found his body a fortnight later and it was clear that the cycling cape he had been wearing at the time had funnelled over his head when he was in the water, and he was in effect smothered. While it was later reported in the newspapers, there was no question of his family seeking someone to blame. How different the world is fifty years on. I now help the same London-based society organize its activities, but the prevailing blame-culture, encouraged by the media, makes it increasingly difficult for a charity like ours to find the volunteers to lead challenging wilderness expeditions for young people.

I still have a cutting from the Montreal Star dated 14th September 1954. The expedition was over and we were back enjoying the comforts of civilization before catching our flight to London. The photograph shows three boys sitting on piles of baggage reading letters from home for the first time in six weeks. Amusingly, I'm one of the boys, but the letter I'm reading was a bogus one, thrust into my hand by the photographer. Before leaving for the expedition I'd told my parents emphatically that they should not write to me – I daresay I felt that it would be ungrown-up and I wanted to cut the apron strings. I regretted it of course as soon as we reached Montreal when all my colleagues tore open their own letters and devoured news from home.

The 1954 expedition was tough, but it was the most unforgettable experience of my life up until then. My parents picked me up in London when we returned in early September, telling me later that it took a further month to rid the inside of their little Ford car of the smell of woodsmoke and stale sweat.

Chapter Three

I enrolled as a fresher at University College London in October, bought my blue and purple college scarf, and began my three-year degree course. Geology was, of course, my main subject, and I took ancillary zoology and geography. Dan Reekie had advised me to include chemistry as an ancillary subject (probably reflecting his own mineralogical background) but time-table problems prevented me from doing so. Bearing in mind my failure to pass chemistry at A-level, this is something for which I thank the gods. Most of my fellow students were in halls of residence or in digs, but by now my ironmonger's part-time job was earning me fifteen shillings a day and I was reluctant to give it up, and so I lived at home and travelled in daily on the 8.05 train from Harold Wood.

With two years' experience of the ironmonger's trade I felt I ought to be able to find a shop nearer home in which to work, and preferably where the management would be more congenial. I found one in Romford. It was an old-established, traditional, ironmongery firm owned by the Smiths, a kindly father and son. I remained there for my three years at UCL, working on Saturdays and part of every vacation. By the time I left I was earning 25 shillings a day and was proud of the knowledge I'd acquired and the service I gave to customers. Such shops no longer exist. We counted out screws by the dozen, sold nails, whiting and shellac by the pound, and measured out garnet and button polish from heavy earthenware jars kept down in the cellar. The wallpaper we stocked had narrow paper selvedges and one of my jobs was to operate the machine which trimmed these off, thus saving the customer from having to do so at home with scissors. It was a hand-powered device and involved cranking the paper through the machine with one hand, while controlling the track of the revolving blades with the other, a similar skill to the childrens' party trick of patting the top of your head while stroking your chest. It was a skill I never fully mastered, as my customers must have discovered from time to time. It still surprises me that no one ever returned or complained about the rolls of wavy-edged paper I sold.

My colleagues at Smith and Son were career ironmongers. They included Mr Davies, a kind and mild-mannered man with a down-trodden air, and Reg, the manager. In the shop's quieter moments, when we were able to enjoy a quick cup of tea (which it was my job to bring in a large enamel jug from the café nearby), Reg would give the staff another update on his domestic life. In particular, Reg would dwell on his perplexing inability to satisfy his wife sexually – adult talk which the virgin that I was found fascinating.

The Canada expedition had stimulated my passion to see more of the world, and in the summer of 1955 Jim Coates and I made a hitch-hiking trip to France. This was my first visit to Europe. We reached Paris by rail and ferry, and what a backward place it seemed. The buses were ancient and had open backs, and the public lavatories were of the flush-and-jump-clear type. With a budget of just a few francs per day for food, we lived on bread and fresh fruit and so we were never able to sample the country's fabled cooking. And the makeshift youth hostel was the space beneath the grandstands of the Malakoff Stadium. After Paris we decided to separate, to improve our chance of getting a lift, as hitching was more difficult than in England. We took up positions a few hundred yards apart beside the road at Fontainebleau, and in no time we were travelling again. I saw Jim climb into a car with a group of young guys and head off to the South, and soon after I was seated beside a pretty girl, practicing my French as we headed toward the Jura. When I got back to Paris I arrived in time to catch Jim setting out to spend his last few francs on an evening at the Folies Bergère. I have been back to Paris many times since that first visit, but, while I admire it visually I have never grown to love the city as others seem to.

Jim and I had gone through school together, although he had left before the sixth form, and it was he who introduced me to sailing. Together we would charter a 14ft sailing dinghy, the *Jolly Roger*, from a Maldon boatyard and spend two or three days at a time exploring the muddy creeks of the Blackwater estuary, sleeping uncomfortably under the thwarts and cooking our meals on a paraffin stove. Harnessing the wind to power our small clinker-built craft through the muddy water was sheer pleasure, and sowed in me a lasting love of cruising in coastal waters. Meanwhile Jim went on to become a successful racing yachtsman.

The head of the Department of Geology at UCL was Professor S.E.Hollingworth. Lecturing was not his forte – in truth he was a disorganised mumbler in front of a class of students. But what he may have lacked as a lecturer he made up for as a teacher in the field. Before geology became dominated by mathematical modelling and computer simulation, geology was still a science which relied on accurate observation and interpretation in the field. Standing on a grassy hillside with a group of students there were few as good as Hollingworth at tracing where a particular sandstone bed might underlie a slight break in the slope, or where a fault may be present to account for an offset in the topography. Our first year's field mapping course was in the Carboniferous Limestone hills of the Mendips. Every evening at the pub in Shepton Mallet where we lodged and boozed on scrumpy, we would plot the day's observations onto our maps, watching the geological picture gradually take shape as if it were a jigsaw puzzle.

There were about eight of us doing what was called Special Geology, all male. I say male because we were hardly yet men. In that limbo between adolescence and adulthood I did all I could to appear grown-up – I had taken to wearing a trilby hat and smoking a pipe, and I even grew a wispy beard on one of the field trips. I had also changed my nickname from Mick to Mike, thinking the latter sounded more manly.

Although the era of women geologists had not yet arrived, the geography and zoology courses I attended were more evenly split between the sexes. Mixing with young women was something I slowly got used to, and self-consciously began shyly to enjoy.

My geography course lasted just one year. At school I had been taught by Harry Askew, a one-time Olympic long-jumper and an inspiring teacher, and geography had been a favourite subject. But now I was required to endure whole afternoons learning tedious things like graphs and pie-charts, whereas I wanted to hear about the Himalayas and the Amazon. Professor Darby, Eric Brown, Coppock – I recall their names, but little of what they taught us. I remember best the field trip, a week in the Cotswolds based at Cowley Manor making land-use maps and speculating on the origin of farmland ridge-and-furrow. Zoology was much more engrossing, particularly the comprehensive course which tackled the vertebrates in the first year and invertebrates the next. The Department boasted many famous names in the world of zoology, including Peter Medawar who taught embryology and had us grafting wing buds on unhatched chicks, Pamela Robinson the vertebrate palaeontologist, and Gruneberg who lectured us in his thick German accent on "ze genetics of ze fruit fly."

In the summer of 1956 we undertook our main mapping project, the results from which would be submitted the following year to count toward our degree. I found myself allocated an area in Arctic Norway, on the edge of the Svartisen ice-cap. I went with three other students, two of them post-graduates who were working toward their PhDs, and the other a fellow undergraduate, Mike Holmes. It was my third successive summer abroad and, again, it left a lasting impression. We sailed from Newcastle to Bergen and then took the coastal steamer which called at every small town as far as our destination, Glomfjord. Before the present network of inland roads was built, the *hurtigrute,* as it was called, was the lifeline of the coastal settlements, and its arrival at bustling jetties was an important event – amid, I remember, the ever-present smell of fish.

We lived that summer in a wooden hut on the shore of a large upland lake, Storglomvann. Eastwards the plateau rolled on toward the distant Swedish border, but to the south the view was of the ice-cap and its glaciers, calving icebergs which floated off across the lake. The terrain was bare and the metamorphic rocks were marbles and garnet schists whose swirling outcrop pattern stood out clearly, leaving little for us to interpret. Western Norway has frequent rainy days, but when it was sunny we would row out among the icebergs or fish for trout to supplement our rations. But the mosquitoes were a constant torment, no matter how I puffed on my tobacco pipe hoping to repel them.

When our project was complete, Mike and I chose to return to England the long way. We sailed further north to Narvik where we took the train east, our carriage full of Swedes returning home, surprisingly, with crates of margarine. Past the iron-mining town of Kiruna and into Finland we rode, winding our way through its thousands of lakes to Helsinki. A group of Russians led by Molotov, their Minister of Foreign Affairs, was leaving Helsinki station with much ceremony as we arrived, and we were reminded that parts of the country were still under Soviet control. (I was too young at the time to appreciate the irony which connects his name with the petrol bomb beloved by terrorists and freedom-fighters, as it was the Finns who coined the name 'Molotov Cocktail' during their resistance to the Soviet invasion of their country – the so-called Winter War – in the early 'forties.)

We took the night-time crossing to Stockholm, dancing on deck with any young Swedish girl willing to put up with us in our less-than-trendy field gear and boots. Our travel budget on this trip was so tight that when we finally arrived by ferry at Harwich I had no money left for the rail fare home to Harold Wood, and so was forced to hitch hike.

It was about this time that John and I acquired our first car, a 1935 Morris four-door saloon. Determined that we should be distracted from the dangers of motorcycling, my father had bought this ancient car for us for twenty pounds, having first satisfied himself that it was mechanically sound. We shared it of course, John taking it for a few weeks at a time to his airforce camp (as by now he was a National Serviceman) leaving me to use it the rest of the time. It was a heavy, four-square, softly-sprung, matronly sort of car with worn leather seats, and it drove at a sedate thirty or forty miles per hour, or maybe fifty when pressed. My County Major Exhibition grant from Essex was about £150 per year, on top of which I had my part-time earnings as an ironmonger, and so with petrol at only a couple of shillings per gallon I was just able to cover the running costs. One thing was certain, it opened up social opportunities in a way that a bike was never able.

One Saturday night in March of my final year at university, I went to the local dance hall in Brentwood. Dressed in my best suit – my only suit – and with hair slicked down with Brylcreem, I joined the chaps at the end of the hall and surveyed the girls seated around the edge of the dance floor. The band was playing Guy Mitchell favourites as I asked a pretty, dark-haired, girl if she would care to dance. Neither of us were good dancers, and I remember how nervously and stiffly she held me at arm's-length. But at the end of the dance I took her home in the Morris and we agreed to meet again. Her name was Anne Gallagher, a primary-school teacher who lived in nearby Hornchurch with, I discovered later, her large and devout Catholic family.

At university I acted the fool too much. In one palaeontology practical class, for example, I was asked to leave when the lecturer decided that entertaining my classmates on the classical guitar at the back of the room was a step too far. But geology remained a passion and I absorbed the subject effortlessly. In our Finals Year we undertook one last mapping project. It was run over the Easter Vacation and we worked from a guest house near Coniston – a memorable trip since I was charged with the responsible job of driving the Department's Landrover there and back.

In the months leading up to our final exams the petrology lecturer Reg Bradshaw set me weekly essays, recommending that I should write them as if I were under examination conditions. This proved invaluable training as, no matter how obscure the essay topic, it taught me how to muster everything I knew and put it down on paper. When the examination results were announced I was surprised to find I'd been awarded a First Class Honours degree.

Through the summer of 1957 I continued working at the ironmongers shop but found the time to take Anne on holiday to Cornwall in the Morris. We had another couple with us, Dusty Lewis and his girlfriend, and from a 21st Century perspective the innocence of that holiday – with boys in one room and girls in the other – is almost unimaginable.

Investigating job opportunities began to loom important as the summer advanced. The Colonial Geological Survey was recruiting for its Ugandan office, and the prospect of working in Africa was not unattractive, so I was pleased when an offer of the position of field geologist arrived by post. Enclosed with the offer-letter was a slim book in a special insect-proof binding on how to deal with a formidable catalogue of tropical diseases. But what disturbed me more than the possibility of catching typhoid or elephantiasis was the statement that I should need to buy a Landrover. Having driven the University's Landrover on the Lake District field trip I had developed a liking for these vehicles, but the news that to do my job in Uganda I should have to buy one out of my own money struck me as a poor deal. After further discussions with the Colonial Office in which I singularly failed to overturn Government policy, we decided to go our separate ways. In any case, I was now seriously courting Anne and so had growing misgivings about leaving her for darkest Africa.

Wimpey, the engineering contractor, offered me a job testing soil samples in their Hounslow laboratory. It would have meant I could remain in England, but it would have been a boring indoors job and not what I wanted. I then tried Costain, a contractor which my "Opportunities for Graduates" booklet said was building pipelines in the Middle East. At the interview the personnel manager explained that they had no opportunities for geologists that year, but wondered why I was not applying to any oil companies. The reason I hadn't was that it had never occurred to me to do so; at university I had specialised in so-called hard-rock geology – igneous and metamorphic rocks – whereas, surely, I thought, oil companies would only be interested in soft-rock geologists. "Listen," he said, "the personnel chap at BP is an old friend of mine. If you like, I'll give him a call." There and then he got him on the telephone: "I've got this young geologist in my office; he's got a good degree and seems a bright sort of lad. What do you say I send him over?"

A week later I was sitting in Britannic House in Finsbury Circus, BP's head office. Once Personnel Department had finished with me I was sent to meet the Chief Geologist, Norman Falcon. A lean and craggy-featured man from years of fieldwork in Persia, he inspired immediate admiration and affection. Through his considerable stammer, which embarrassed me more than it did him, he delved into my geological knowledge and we talked of travel and of expeditions. A week passed before the hoped-for letter arrived with the tell-tale BP shield on the envelope. I tore it open to find that I was being offered the position of Geologist, on a salary of £1075, and I was to report to BP's Midlands office on the 9th September. I accepted without hesitation.

The train from King's Cross dropped me at Newark where I was met by a chauffeur in a black Austin Princess who drove me to the small Nottinghamshire town of Southwell, where the Company had booked me into the Admiral Rodney Hotel. The next day I was picked up again by car and taken to the base for BP's UK operations – a collection of green-painted huts and workshops surrounded by rolling countryside at the edge of the sleepy village of Eakring. My new boss was Ken Roberts, a friendly and welcoming Australian geologist who was in charge of BP's search for new oilfields in Britain. A drilling campaign in the 'thirties and through the war years had found several small oilfields beneath the woods and meadows of Nottinghamshire. Although they were tiny by world standards, the tax regime in the UK favoured domestically-produced oil, which meant they were profitable and the risky business of looking for more of them could be justified.

Within a week I was appointed resident geologist on an exploration well being drilled at Askern, north of Doncaster. It was a typical drilling site: a bleak bulldozed area of mud and hardcore half the size of a football pitch, with a scatter of huts, pipe-racks, tanks and, at the centre, a steel derrick rising above immense diesel engines and clattering machinery. The geologist's laboratory was an over-heated caravan where Bert, the assistant, washed and laid out the drill cuttings and made the tea – or "mashed the tea" as Bert would say, as these were Nottinghamshire men with their own dialect.

The geologist I was replacing was Roland, a young Irishman with a crazy streak and an exaggerated Australian accent picked up during his recent two-year posting on drilling rigs in the jungles of Papua New Guinea. The hand-over was completed in a day, and that was the extent of the training I was to receive for what seemed to me a position of some responsibility. As the drill-bit ground its way toward its target a mile or so below the surface, cuttings were recovered at frequent intervals and I examined them under the microscope. I would write up my observations in reports to headquarters and plot the strata in the form of a log, hoping to predict when a particular target zone might be encountered. It was then up to me to instruct the drillers to stop and take a core, a continuous, thin, cylindrical sample of rock. If the core was encouraging and wept oil or gas we had to establish whether it would flow to the surface, a potentially hazardous operation as if it went wrong it could cover the well-site with mud or oil, or worse still, we could have a gas blowout. But thankfully such mishaps were rare.

I soon got the hang of things, picked up the drillers' jargon, drove around in one of the Company's green Austin vans, and began to enjoy this grown-up world of the oil geologist. The borehole at Askern turned out to be a failure and so it was filled with concrete and the drilling rig moved on. But my next one, at Calow, was a gas discovery. I stayed at the Portland Hotel in nearby Chesterfield, a town noted for its crooked church spire. It was a cold and wet autumn and after dining in the hotel grill-room I'd pass the evenings reading or writing in my dreary bedroom, feeding the electric heater with shillings until it was time for bed. Drilling went on around the clock for seven days a week, but whenever I could I'd drive to Essex to spend time with Anne. By then we were engaged, undeterred by my father's growing displeasure over my affection for a 'papist'. In this age of co-habitation and routine pre-marital sex, the notion of an educated twenty-one-year-old male entering into marriage is almost unthinkable, but my instincts were stronger than my reasoning powers.

The Great North Road from the Midlands to the South was a test of fortitude and navigation. It wound its way through villages and town centres, with only rare stretches of dual-carriageway. The supposedly improved sections were three-lane stretches where overtaking in the centre lane was a free-for-all. Grantham was about the half-way point, and I remember one occasion when I broke my journey there for a meal. It was evening, I parked the green Austin van in the courtyard of the historic coaching inn, the Angel, and found a table in the dining room. The waiter brought me a wine list with the menu, and to accompany my steak and kidney pie I summoned as much savoir-faire as I could and ordered a half-bottle of burgundy. It was a Beaune. I remember it clearly because it was the first time in my life I had ordered wine with a meal – a small milestone on my road to adulthood.

I had agreed with John that I would take over the Morris for a spell, as the Austin van was being assigned to someone else. I missed the speed and reliability of the van – more than twenty years younger than the Morris – but the mileage allowance I could claim from BP was some compensation. I had moved on from the Calow gas well and was working now at a drilling site on the high moors a few miles outside Whitby. A snowy winter had descended, and for days the temperature remained below freezing. Driving up the steep hill onto the moors became a serious challenge, which could easily end with the car sliding into a ditch. And our problems weren't over once we'd reached the rig site, as we then had to thaw the frozen water pipes to the laboratory with diesel-soaked rags to get it working again. Only then could Bert begin the job of mashing jugs of hot tea.

But those harsh winter conditions were not to the old car's liking. One morning, after trudging through snow from the George Hotel in Whitby to the town-centre carpark, I settled into the driver's seat, switched on the ignition and was startled to see smoke rise from under the dashboard. I peered beneath and could see red-hot wires, so gingerly put in a gloved hand, grabbed them, and pulled, fearing that if I did nothing the car would soon catch fire. A tangle of copper and melted insulation came away in my hand. I dropped it sizzling onto the snow. When I turned the key again, to my surprise the engine started. I drove out of town and up the snowy hill to the rig site, but she was not the car she had been. The circuitry had developed a serious flaw. For the rest of her life other motorists would wave and point, signalling that I was driving with my headlights on. They little knew that if I switched off the headlights the engine of our poor old Morris would splutter to a standstill. She finally gave up the ghost a year or so later as John was driving home across Devon from his RAF camp. We sold her to a scrap-yard for £5.

Chapter Four

It was while I was working on the well being drilled outside Whitby that I got news that the Company planned to post me to Sicily. I was happy enough working in England but the main attraction of joining BP had been the prospect of living and working abroad – Anne notwithstanding. And so with thoughts of Mediterranean sunshine I set off for the last time to the headquarters office at Eakring. After making my farewells, and with the airline ticket in my pocket, I was stepping in to my car to drive off to Essex and then to the airport when Ken Roberts, my boss, called out from his office: "The Chief Geologist's on the phone from London," he shouted, "the well in Sicily has struck water and has stopped drilling. He wants you to go back to Whitby!"

A few weeks later, news of another posting arrived – I was being sent to an equally sunny Mediterranean country, Libya. Even if the Sicily trip had come off, it would have been only a temporary sojourn there. But Libya was to be a full-scale posting. A kit-list arrived from Personnel Department and I bought the recommended items from Alkit, a supplier of tropical wear in Piccadilly. I recall the list included a lightweight white linen suit which – neophyte that I was – I didn't question; I eventually gave it to a charity shop many years later, having never worn it. Anyway, this and all the other items were sent on ahead by air freight in the recommended insect-proof steel trunk – an item which by contrast I used countless times and have to this day.

Nothing intervened to derail my posting this time, and on a grey February day in 1958 I left London. It was not a joyful departure. Anne and I had little idea when we would see each other again, and my father was tight-lipped still over the whole business of his young son's engagement.

I was not yet a seasoned flier and was puzzled by the indifference of other passengers to the unfolding scenery below. It was a piston-engined plane of course, an "Elizabethan" I think it was called. As we droned south across Europe the passengers dozed, or enjoyed their BEA meals, while I drank in the view of the snow-covered Alps and the Cote d'Azur (for we stopped to refuel at Nice) before our descent over the brown sandy coastal strip of Libya. I had never known such intense sunshine as struck when I stepped from the flight at Castel Benito Airport – and yet it was still only winter. A driver greeted me in halting English as I emerged from the customs shed, and I climbed into his Company Landrover. We rattled through a dusty landscape of orange and olive groves which seemed to struggle to keep the drifting sand at bay. Shabby, white-washed, mud-brick houses became more numerous as we approached the city, and sheep and camels grazed among the roadside prickly-pear hedges.

The Del Mehari, one of only a couple of hotels in Tripoli at that time, was rumoured to have been Rommel's wartime headquarters. Its bourgainvillea-hung balconies and berber architecture belied its cell-like bedrooms, and this was to be my home for the next few weeks. Winter gave way to spring and the weather became hotter. At night the room was stifling, and through the open window the sound of wailing pye-dogs filled the air till the muezzins' calls from nearby minarets signalled the dawn. It was said that the only air-conditioners in the entire country were in the nearby American air-base at Wheelus, (although that changed as oil companies introduced them to their larger desert camps).

Before the overthrow of King Idris by Colonel Gaddafi some eleven years later, Tripoli was an Arab city but with a strongly Continental flavour. A large Italian population lived unmolested and were free to carry on their businesses and to worship in the city's cathedral. Of an evening we would often stop to sip a beer at one of the cafés on Sharia Istiqlal while Italian families in their best clothes processed to and fro, our eyes firmly on their pretty daughters. A number of these nubile creatures were employed as local staff in our offices, but were well known to be as unattainable as goddesses.

In those first few weeks I made several short trips south from the coast, including one to the rig which was drilling an exploration well for BP near the airport. The resident geologist was Don Sheridan, a quietly-spoken Irishman who had spent several years in Iraq before coming to Libya, and so in my eyes was something of an 'old hand'. Only a small percentage of exploration wells that are drilled worldwide turn out to be successful, and all the signs pointed to this being another "dry hole." We sat in his caravan as the rig clattered nearby, talking about geology and trying to reconcile what the well was finding with what had been forecast. As the afternoon drew on, Don asked if I should like some tea. He called out to Ali, his houseboy: *"Ali, jeeb chahi min fudluk."* A brief conversation in Arabic followed, and I listened with growing admiration.

I now realise that it's easy to impress others with even a smattering of a foreign language, and I don't know if Don's conversation represented more than just a smattering. But impressed I certainly was. If one of my colleagues could have such a grasp of Arabic, I decided, then I too would try and learn the language. At school I had shown no particular aptitude for French, and even less for Spanish. But now I could see how satisfying it would be if, like Don, I could speak with the local people in their own language. Before leaving Libya two years later I could get by in Arabic. And in the years that followed I acquired a basic grasp of Thai, Malay and Russian, as well as improving on my schoolboy French and Spanish. Japanese, I'm now finding, is coming more slowly.

The Libyan oil industry was born that year. A number of western oil companies including BP had been granted concessions over what were thought to be the more promising parts of the country. Drilling activity was picking up and major oil discoveries were being made. One of BP's concessions was a vast area of barren limestone hills and black basalt bouldery desert south of the oasis of Hun, about 300 km from the coast. I was assigned to a drilling rig whose job was to find not oil but water in this concession. The rationale was that if deep drilling were to be undertaken at a later date, then an abundant supply of water for the operation would be vital.

The Davy Camp, as it was called, was to be under the leadership of a tool-pusher named Pat Gibbings, a public-school-educated man who I remember particularly for the idiosyncratic Italian with which he addressed the labourers. He had picked it up while working on BP's abortive Sicilian drilling venture and it was peppered with *"adésso; òggi non domani!"* to inject urgency into all of his orders. Beneath Pat was a team of three drillers who, like myself, had recently arrived from rigs in Northern England. And I was to be the geologist.

First, we had to get the rig from the yard in Tripoli down to the site in the desert where it was thought that water might be present at depth. It was a truck-mounted rig and weighed over forty tons, and, once away from the coast the road south was no more than a corrugated and potholed track. We took turns to drive, finding that going across the desert was mostly smoother than the track. Any attempt to lift the speed above about 20 kph was punished sooner or later by crashing into a pot-hole, which threw us hard against the roof of the cab. We completed the 800 kilometre journey without any serious breakages, and arrived with our small convoy three days later, bruised and dust-covered .

The broad wadi where the well was to be drilled was flanked by flat-topped limestone hills and was barren but for a few brittle, prickly, shrubs clinging to life along the valley bottom. There was a caravan for each of the Europeans, and a line of tents for the several scores of labour. Over the next few months the heat became merciless and inescapable. Dust-devils danced through the wadi as the thermometer climbed, and we experienced one particularly fierce sandstorm. All of that day small puffs of cumulus had spread across the normally cloudless sky until, in the distance, we could see a dense grey mass, a flattened ball a half-mile wide that hugged the ground as it approached our camp. Minutes later we were engulfed in a swirling maelstrom of dust and sand which rocked our caravans and sent the tents flying. Empty oil-drums were overturned like skittles. Lightning lit up the gloom and the thunder crashed. Before the storm moved on the heavens opened and the camp was deluged with red muddy rain.

Drilling progress was slow and beset by mechanical break-downs and stuck-pipe problems, which meant there was little need for a geologist. To keep myself occupied I took on the job of relief tanker-driver, fetching water for our camp and for our drilling needs from a distant oasis. I enjoyed the navigational challenge it presented, particularly at night when the moon and stars offered the only way of staying on course through that wilderness of gravel and boulder-strewn desert.

After a hard day working in the heat of the sun on the rig floor, drillers like to relax with a glass of beer, and I was not averse to joining them. My favourite drinking partner was Barry, a grimy driller in appearance but with a sensitive side to his nature. He was a Shrewsbury School man and had brought with him into the desert a small collection of gramophone records and a player. Relaxing over a cold beer in his caravan, with door and windows thrown wide to disperse the heat of the day, we would send symphonies and concertos rolling into the star-lit desert nights. The more we repeated them, the more I grew fond of them: Wagner's *Siegfrid Idyll*, *Romeo and Juliet* by Tchaikovsky and, my favourite still, Sibelius' *E minor First Symphony*.

Barry was devoted to his wife Marion, living back in Tripoli with their young child. Where other men might have kept a photograph of their loved one, Barry drew comfort from a pair of his wife's silk panties. I doubt if he took them with him to the drilling-rig every day – I'm not suggesting Barry had a panties fetish – but in the caravan of an evening Marion's panties were always close to hand and were even put to use when a record needed cleaning before placing it on the turntable.

Since leaving home, Anne and I had continued our courtship by letter and we agreed that I should return to Essex in the autumn to marry. Management in Tripoli had made it clear that I could have no home-leave until my two years were up, and as for married accommodation, it was Company policy not to make this available to young men on their first overseas posting. We were, however, entitled to field-breaks back in Tripoli after six or seven weeks in the desert. The breaks were usually of about a week and I resolved that, notwithstanding the rules, I would fly to England in November on my field-break, marry, and come back to Tripoli with my new wife.

This I duly did, and the Company had little choice but to provide us with a flat. It was a modern white stucco apartment on a dusty, unsurfaced, road at the edge of town. From the first-floor balcony we could watch the comings and goings of sheep and camels and the local people, the women enveloped in their cream-coloured *barakan* from which a single eye would peer, and the men in their trademark red felt *tarboosh*. Beyond an area of open ground a white-painted, mud-brick, domed building stood by itself, and sometimes of an evening strange noises would come from it. We would hear rhythmic drumming and sometimes wailing as we relaxed on our balcony, and we learned later that these were *marabout,* holy men. Into this scene from Barbary a continental note was injected by the cry of the Italian vegetable-seller. With his barrow piled high with produce he would progress along the pot-holed road, calling out *'carciófi ..... carciófi .... meloni!'*

Life for foreigners in Tripoli was generally safe, although hostility often seemed close to the surface, perhaps not surprisingly bearing in mind the Libyans' history of colonial subjugation. Friends spoke of having been spat upon while walking through the narrow alleys of the old city, and there were reports of cars being intentionally damaged. In July of that year, 1958, I was enjoying a field-break in Tripoli when news came of the revolution in Iraq and the murder of King Faisal II. Rioting broke out and for a few days it was safest to remain indoors to avoid the chanting crowds gathered in the main streets.

Relations between the races were not helped by the overbearing manner of many of the expatriates, and as a callow twenty-two-year-old I confess that I myself was not always a shining example of tolerance and understanding. While I admired and enjoyed the company of the desert people, I found the Libyans of Tripoli and of the coastal region less congenial, as they often showed a sullen dislike of foreigners. Without exception the Libyans with whom we had contact were carrying out unskilled or semi-skilled jobs. Looking back, I now see that one of the factors contributing to the resentment of Europeans by the Libyan may have been the near monopoly of 'the professions' by immigrant Italians, meaning that a Libyan middle-class was virtually non-existent.

I recall attending a reception in Tripoli to welcome a newly-arrived senior manager who was to be our new boss. A wise old bird, he had worked in many parts of the world including a period in Persia before the war. In the course of talking with him I commented on how irritating the local people could be. In a salutary slap-down that I fully deserved, he gently pointed out that we were guests in their country, and that if we found them difficult to relate to perhaps the fault lay as much in ourselves as in our hosts.

It could not have been an easy life for a young bride. Anne had no more experience than I of such an exotic culture, and I was away from home in the desert for long periods. Although she soon made friends with other expatriate wives, the hierarchy among the BP womenfolk – reminiscent of stories of the Raj – was not something she was used to. Coffee-mornings were a prominent feature of their social life, and although they were informal and enjoyable if held among friends, when they were hosted by the boss's wife they were anything but informal. The top BP manager, Macpherson, was a kindly man who had spent his earlier career in the Middle East and was pleased to lend me books to help me with my Arabic studies. But his wife was of the old school, and her coffee-mornings meant wives donning their best clothes, and rising to their feet when she entered the room.

Early in January the following year I teamed up with geologist Joe Glance to carry out a major survey in the Fezzan, Libya's southwestern province. Our orders were to examine the surface geology of the region to assist the interpretation of data which was then being gathered by BP seismic surveys in our Fezzan concession. There were seven in our party: Joe and I, our two field-assistants, two drivers, and Sadek our cook-cum-houseboy. We geologists each had a Landrover, and our camp equipment, fuel, food and water were carried on a Bedford 4x4 and a Dodge Powerwagon.

Much of the Fezzan is covered by sand-seas. Each of these is two- to three-hundred kilometres across: vast tracts of inhospitable billowing dunes through which it is difficult to travel, and which in any case offer little for the geologist. But around and between the sand-seas are areas of outcropping rocks, and it was on these that Joe and I concentrated our attention.

Joe was a crusty introvert who had little time for young geologists, in his opinion still wet behind the ears. But his taciturnity didn't prevent me from enjoying the next six months. This, after all, was the sort of work for which I had joined BP.

Everything in base-camp bore the mark of Benjamin Edgington Ltd, the company that had equipped generations of empire-builders. Joe and I each had one of their "Swiss Cottage" tents, with an enclosed porch at one end where the folding wash-stand was set up. We ate and wrote at folding canvas-topped tables, and our beds were folding camp beds. And then there was a cook-tent for Sadek and another tent for the Libyan personnel. Our personal gear we stored in *yakdans*, canvas and rawhide chests tough enough to withstand being carried by camel, and left over from the days when our forerunner, Anglo-Iranian, was exploring in Persia.

I had never imagined the desert as anything but hot, but in January we would often wake to find ice in the water buckets. After breakfasting on Sadek's freshly-baked bread we would set off for the day's geology, muffled up against the cold. Joe and I each drove our own Landrover, Joe invariably leading the way with me behind, choking in his dust cloud. When we came across an oasis we would stop and buy whatever fresh food they might have: eggs, dates, or maybe a sheep or a goat which would get its throat slit that evening when we returned to camp. The desert people showed none of the unfriendliness that one came across on the coast, and whether it was a group of blue-swathed Tuareg atop their camels or a stall-holder in some *souq*, we would often be invited to join them in a glass of their strong, sweet tea, brewed in an enamel kettle over an improvised wood fire.

We roamed in our Landrovers over great distances, often fly-camping under the stars when it became impracticable to return for the night to base-camp. At times we had no choice but to cross areas of sandy desert and, as often as not, one of us would hit a soft patch and the Landrover would sink to its axles and stop. It meant digging the sand away and pushing sand-ladders beneath the wheels, repeating this yard by yard until back again on firmer ground, a tedious and heart-breaking job. And then there were areas of *feshfesh*, innocent-looking level plains where a surface of fine gravel conceals the finest dust, and getting out of these was even harder than digging out of soft sand.

A daily rhythm developed as winter gave way to spring, broken only by our too-infrequent field-breaks in Tripoli. Sequences of strata would be measured and meticulously logged; fossils would be collected for dispatch to our palaeontologists at Sunbury-on-Thames to help us correlate the rock-successions from place to place; and rocks possibly favourable for oil were sampled for later study in the laboratory. As the days and the weeks went by our understanding of the geology increased and the colours spread across the map.

In the far southwest, on the Algerian border, the Acacus Mountains are a spectacular range and we spent some weeks working there among the towering sandstone pinnacles. They resemble the mountains of Arizona's Monument Valley and some Tuareg had told our men that certain of the mountains were inhabited by *djinn,* spirits that were not necessarily friendly. We came across no *djinn,* but one day we were fortunate to run across an Italian archaeological expedition exploring the area's rock art. Its leader, Fabrizio Mori, was later to become the acknowledged expert on Saharan pre-history, and he guided us to the cliff shelters where he was working. It's a remarkable mountain range and not difficult to imagine it as a habitat of early Man. Among the deep and winding ravines we found rare pools of cool, fresh water, and in places the foot of the cliffs was eroded to form an overhang. In such sandstone shelters ancient hunters had inscribed and painted red-oxide images of themselves and of the animals they hunted: buffalo, giraffe, rhinoceros, crocodile and elephant. And other images were of domesticated animals. Mori's experienced eye was able to identify a variety of artistic styles which indicated to him Man's presence over a long period, a period when the Sahara enjoyed a wetter and far less hostile climate than now.

I was fortunate to make some archaeological finds of my own. Once, while we were working in the Jebel ben Gnema, I was lucky to stumble upon a small slab of red sandstone which Stone Age Man had carved with a pair of beautifully-executed giraffe – an artefact that now graces the top of the piano in our London flat. And another time as I was crouching, attending to a call of nature with my back braced against my Landrover bumper, there on the gravel by my feet was a perfect stone axe-head.

Our topographic maps were small-scale and rudimentary, and we navigated as best we could by dead-reckoning and by a sun-compass mounted on the Landrover's wing. One day we found ourselves by mistake in Algeria, at the gates of a Foreign Legion fort – at Tin Alkoum, I believe. The French can be relied on to live well wherever they are stationed, and we lingered there, being entertained over a bottle of good Bordeaux by the fort's immaculately-turned-out commandant. On another occasion we were working close to the border when our Landrovers were buzzed by a three-engined Dornier airplane. The Algerian crisis was causing great problems for de Gaulle's government, and as the plane swung low to investigate us we could clearly see the squad of heavily-armed military inside. We stood beside our vehicles looking as friendly as we could, and waved nervously. When satisfied that we were not rebels, the plane turned west and disappeared into Algerian airspace.

At basecamp, cottage pie became Sadek's speciality, made from potatoes, tinned corned beef and tinned peas. He was also quite imaginative with the dishes he could create from a tin of sausages. After he had served us dinner and the day's geology had been plotted, we would relax. Joe would sit silently puffing on his pipe, while we listened to the BBC World Service on our short-wave radio – Victor Sylvester and his ballroom orchestra was the programme I remember best.

Another diversion I recall with a smile was one we enjoyed as we explored the wadis on the south side of the Gargaf mountain massif in northern Fezzan. These were narrow and deeply incised dry valleys between high sandstone cliffs. The valleys were mostly floored with expanses of angular boulders bigger than footballs, and finding a path through them, and inching the Landrover over them, was painfully slow; it could take an hour to travel a mere couple of kilometres. In places, though, these boulders would give way to stretches of hard and compact sand where we were able to drive at full speed for a kilometre or so – exhilarating after the enforced periods of snail-like progress. But the real exhilaration was to be had where the winds had smeared a layer of firm sand high up the steep wadi sides. Driving at full speed along the wadi, one Landrover behind the other, we would swing the wheel to climb a short way up the wall of sand. Turn the wheel again and we would swoop down, diagonally across the wadi floor and up the other side, going higher this time. After a few zigzags like this we would be flying high up the wadi sides at a crazy angle, like present-day skateboarders in their half-pipe, the field-assistants beside us crying out in terror as the valley-bottom streaked past below. This would continue, racing flat-out across and up the valley sides, leaving a snake-like trail until the smooth sand gave way again to boulders, and we could fly like birds no more. Even Joe, normally so dour, enjoyed the sport as much as I. It was hair-raising but exciting, and I daresay that if we had hit a patch of soft sand while up there at a dizzy angle we would have lost speed and rolled to the wadi floor. But young men have notions they are indestructible.

Occasionally we would see in the distance a group of gazelle or more rarely a barbary sheep, their survival on such sparse greenery a mystery to us. Then as spring advanced and midday temperatures climbed, the smaller creatures would venture from their winter hiding places. Lizards of ugly spinyness called *dhub*, and fast-running monitors they called *worral* would sun themselves on rocks, and vipers left their side-winding tracks to warn us to be wary. At night, camel spiders the size of your thumb, called solifugids, would appear on the ceiling of the tent, and the sound of scampering jerboas could be heard beneath the bed. Here I must admit to a lifelong abhorrence of spiders, or at least of big ones. Snakes are fine (although I keep poisonous ones at a safe distance). Scorpions are alright too. Jerboas and mice are nice. But spiders, if they are big, hairy, and fast runners, make me recoil however harmless they may be. And so it became Sadek's job each night to come at my call and remove the latest spider from my tent before I could settle down to sleep.

Deep in the desert, birds were fairly uncommon, but we did sometimes come across a flock of sand grouse, silly birds and so reluctant to take flight that we were tempted to try and catch one for the pot. The nearest we came to succeeding was when we took our emergency Very pistol, loaded the cartridge with chopped bits of lead in place of the flare, and fired it at our prey – but a Very pistol is no substitute for a shotgun and the birds ran off. Great bustard too, the heaviest of all flying land birds, would run ahead of our vehicles and magnificently launch into the air and fly a hundred yards before running again. But one of my most delightful ornithological memories is of the migrating warblers which would settle in our camp, exhausted, and rest within inches of us while they recovered their strength before flying on.

With the growing heat our need for drinking water grew, and the canvas water-bags tied to the Landrover's side would be emptied by noon. We would replenish our supplies at any oasis we came to, often just a primitive well where we would jostle with villagers and their animals to fill our jerry-cans. We probably knew that to drink this without first purifying or boiling it was foolhardy, but convenience outwieighed caution. It was not long before I became ill with amoebic dysentery.

The heat became a problem too for our transport. Driving over hot and jagged rocks played havoc with ordinary tyres and we found we were having to replace one or two per month. This continued until BP's Stores Department in Tripoli started supplying us with Michelin radial tyres, which had the strength to survive the rough terrain. Another problem which grew with the advancing summer was that petrol in the fuel pumps would vaporize and bring vehicles to a standstill. Fortunately, Landrovers were simpler in those days and we carried out our own repairs and maintenance, discovering that this vaporization problem could be overcome by plumbing-in a second electric pump in parallel with the existing one. But the fuel pump on the Dodge is mechanical, not electric, and one very hot day when far from camp the engine spluttered and stopped. We knew there must be bubbles in the fuel pump, but in removing it, the aluminium casting broke. This was a serious setback, as all of the spare parts were back at camp. After a lot of head-scratching we hit on a possible makeshift remedy. We decided on dispensing with the pump and relying on gravity-feed to get fuel to the carburettor. The sun was setting as we limped back into camp, and a cheer went up for Hussain the field-assistant, gamely sitting on the roof of the cab and ladling petrol into a funnel stuck into the end of a length of hosepipe.

By June the survey was complete. We had mapped the rocks outcropping around the entire rim of the Ubari Sand-Sea, and in a half-circle around the northern half of the Murzuk Sand Sea. It was time to head back to Tripoli for the remainder of the summer. Having had more than enough of driving over the previous five months, Joe and I managed to hitch a ride on a plane from the Fezzan capital, the dusty little town of Sebha. Those flights in old Dakotas could be miserable. In the heat of the day they lurched through the turbulent desert air, often with the overpowering smell of stale vegetables remaining from some earlier flight – they were a test for even the strongest stomachs, let alone one as delicate as mine had become. Before boarding our plane we left clear instructions with the rest of the team, including the two additional drivers sent down from Tripoli, on the route they should follow to get back safely.

It was good to be home, enjoying good food, showers, and a comfortable bed free of wind-blown sand. After a couple of days our convoy rolled into Tripoli, but with only three out of the four vehicles – the Dodge, the Bedford and Joe's Landrover. My Landrover was missing. We quizzed the men and it appeared that my Landrover, with one of the replacement drivers and our field-assistant Ali on board, had last been seen heading away from the convoy in a cloud of dust – it was thought they were probably chasing a gazelle. This was worrying as it was impossible to imagine Ali ever becoming lost. Ali ben Khalifa was a wiry old guy with a wicked toothless grin, who had spent most of his life deep in the desert with the camel caravans. His sense of direction was uncanny. At noon and with the sun overhead you could stop Ali on some featureless gravel plain and ask him the direction of, say, Tripoli or any other town or small oasis. He would look about, hesitate, and then point, saying *"hecky, hecky,"* meaning "over there." We'd check with map and compass, and he was never more than a point or two out. So if Ali had not got lost, where was he?

We waited another day, and when he still failed to appear a major search was put in hand. The Landrover was last seen in the middle of the Hamada el Hamra, a 50,000 square kilometre gravel plain, and not an easy place to look for a missing vehicle. With a tent and supplies in another Landrover I set off south accompanied by a fellow geologist, Jim Hawkins. Crossing the fertile plain behind Tripoli, climbing the winding mountain road to the top of the Gharian Escarpment, and then heading south on a compass bearing it took a whole day to get to the

Hamada. We camped that night and carried on our search in the morning, staying in touch by radio with a light plane which had been chartered to search from above. Hours passed as we quartered the seemingly endless gravel plain. Then at last the plane spotted a distant white mark on the desert. It circled closer and saw the Landrover, and beside it two men waving their arms. Their ordeal was nearly over and the pilot guided us to them on the ground.

Not surprisingly, when we reached them a half-hour later, they were very pleased to see us – they had been stranded there for about five days, they said, and were down to a cup of water per day. With the nearest thing Ali could manage to a shamefaced expression, he admitted to leaving the convoy to chase a gazelle, intending to catch up with the slow-moving convoy later in the day. But, whereas most of the Hamada is as flat as a billiard table, there are occasional gullies, and he explained that while driving at full speed they had gone into one. A lightly-laden Landrover may have survived the crash, but mine had been fully laden and, what is more, it had half-inch steel armour plating welded beneath it as protection against possible landmines left from the War. When it hit the bed of the gully the two front wheels splayed. They tried driving on but further progress was at walking pace, and like plough-shares the wheels left deep furrows in the desert surface. In just a few kilometres they ran out of fuel. Thankfully they had the sense to stay with the vehicle. They had opened my bed-roll and spread my bed-sheets out on the desert to attract attention, and then they waited. This cool-headedness saved their lives; it was 200 kilometres to the nearest oasis and had they tried to make it on foot they would certainly have perished.

It was a good summer. The office closed in the early afternoons because of the stifling heat, and that left plenty of time for trips to the beach with friends. And at weekends we could go further afield and make sightseeing visits to Leptis Magna and Sabratha, well-preserved Roman towns along the coast.

My first child, Helen Angela, was born at 5.30 a.m. at the BP Nursing Home in the Garden City quarter of Tripoli on Wednesday 19th August. In those days fathers were not encouraged to be present, and I simply did what everyone did, and waited to be called.

It was not many weeks after Helen was born that we were at the beach one day, and I was snorkelling with the spear gun which Anne had bought me for my 23rd birthday. On a diving trip a week earlier I had damaged the spear, and in reassembling it I had omitted to fit the line which is intended to connect the spear to the gun. The omission was to have painful consequences.

On the day in question I was swimming about fifty yards from the shore, my face under water as I breathed through the snorkel, when I saw a large ray resting on the sea floor beneath me. I took careful aim and to my surprise, I hit it, pinning the poor creature to the sand. Because there was no longer a line attached to the spear, I now had the task of swimming down to recover both it and the fish. I duck-dived and swam downwards but, before I could reach the fish my breath gave out and I had to return to the surface. I tried again – still no success. I repeated this several times, getting wearier in the process. Finally, determined not to lose my fish and, more important, my spear, I did manage to swim down to where I could reach out and grab the spear. But, I was – and still am – ignorant of how to equalise the pressure in my ears. As I reached out and took hold of the spear there was a shooting pain in my right ear. I struggled toward the surface, fighting the fish's efforts to stay on the sea bed.

At last, with lungs ready to burst, I gulped in fresh air. But the world around me was reeling in a giddy whirl. My sense of balance had been badly disturbed by the crash in my ear and the inrush of cold water. With the horizon seeming to spin I reached the shore, but only with the greatest difficulty. Any sense of triumph over recovering my spear and the fish was blotted out by my utter exhaustion and the pain in my ear. I collapsed on the sand and lay there for some minutes waiting for the world to stabilise. As soon as I was able to stand up, I was driven home. Next day at the clinic, the Company's Scottish doctor pronounced that my eardrum was burst.

I relate this small drama because it is a possible explanation of the hardness of hearing which afflicted me in later years. Friends will attest to my insistence on sitting on their right at social gatherings, so that my good left ear can give them my full attention.

Ron Walters arrived in Tripoli earlier that year to take up the position of palaeontologist. This was his first overseas posting and, like me, he had disobeyed the rules, married, and presented the company with the task of providing him and Bernice with accommodation. Unlike me, he chose to marry in the British Garrison Church in Tripoli. We had become good friends and I happily accepted when asked to be best man at his wedding.

I'm reminded of one exploit of Ron's which happened before Bernice had arrived in Tripoli. Normally punctual, Ron failed to turn up at the office one day. I asked around and called his lodgings – perhaps he was staying at the Bachelors' Mess, I can't remember – and was told that he was in the British Army Hospital. Concerned, I promptly went to see him, hoping that he was not seriously ill. I found him swathed in bandages and rather crestfallen. He explained that he'd been to a very good party the previous night – that much he could remember clearly. But as for the rest of the night his memory was blurred. It didn't take long to piece together what had happened. On his way home Ron had lost control of his Volkswagen Beetle. It had overturned and thrown him into a bed of prickly pear cactus. His relaxed inebriation perhaps accounted for his lack of broken bones, but he had been painfully speared over much of his body by needle-like cactus spines, some of which he was still pulling out several weeks later.

As summer passed and the heat became less fierce it was time to prepare again for another long spell in the desert. The Landrovers and trucks were loaded afresh and in September I set off with Laurie Horobin to carry out the Tripolitania survey. The terrain was different from the Fezzan. Instead of the high sandstone mountains and the sand-seas, we now found ourselves working among limestone hills and on broad gravel plains including the now familiar Hamada al Hamra.

Laurie was an easier fellow to work with than the introverted Joe. A freckle-faced red-head with a cigarette always between his fingers, he was well-known (at least among the wives) for wearing the baggiest and most horribly revealing shorts – true "empire-builders". We worked together until the following spring.

Again, I was to see places that have remained in my mind's eye to this day. The town of Ghadames, close to the Algerian border, is such a place. Approaching from the east we saw first the cemetery, outside the limits of the town, the graves of men marked by pairs of rough stones and those of women by three – the *djinns* of women needing that extra stone to keep them down, my driver laughed. A few black tents with hobbled camels and dark-faced Berber women over their cooking pots. And then the groves of date palms, their lush heads like exploding green fireworks against the cloudless sky. Each building is a whitewashed cube, but atop each corner the walls rise to a point, and there are domes and arches and lintels of mud-brick latticework. Throughout the small town, turbaned men come and go in a network of vaulted passageways and alleys, closed from the sky and as cool as caves. And there are shady courtyards with water-filled cisterns and pools. It was a delight, and an oasis in every sense. No wonder, I thought, the ancients called this place The Pearl of the Sahara.

I've mentioned the Gharian Escarpment (also called the Jebel Nefusa) which bounds the Tripoli coastal plain. One of my first jobs with Laurie was to carry out a series of traverses up the scarp to see how the geology varies along its length. It was straightforward work and we were pleased to find a fossil reptile bed which would form a useful marker as we correlated the rock-succession from traverse to traverse. But we were hampered by the talus which concealed the strata beneath the lower slopes. We thought we might have found the answer when we arrived one day at an abandoned troglodyte village. The former residents had dug caves in the lower slopes of the *jebel* and these former homes, we thought, would allow us to examine the bedrock behind the talus slope. It was puzzling that such pleasantly cool living-spaces were no longer occupied, but it was our good fortune. In half an hour Laurie and I had finished working in the first of the caves, and although the interior was dark we were able to confirm the nature of the bedrock, as we had hoped.

As we emerged into the sunlight I felt my bare legs itching. I bent to relieve the irritation and found a tick from the cave had climbed onto me, and I quickly picked it off. And then I saw another ....and another. As I looked more closely I saw that my bare legs didn't just have a few ticks on them; I was horrified to see they were also covered in fleas. I had hundreds of fleas on me, each one hurrying upward toward the sanctuary of my shorts. And Laurie was the same, his legs alive with fleas. In a panic we picked them off, but no matter how quickly we worked to rid ourselves of those pesky parasites we could see them jumping back onto us again. The situation was getting desperate and the first of them were already up as far as my underpants. Obviously we needed to stand on something high enough that the fleas couldn't jump back onto us again. Of course, our vehicles! "Quick, Laurie, on to the Landrovers," I shouted. In a flash we were up on the cab roofs, Laurie on his and me on mine, each tearing off his shorts and underpants and frantically picking off the fleas and dropping them one by one onto the ground. It took perhaps twenty minutes before the situation was back under control.

Relieved at having more or less debugged ourselves we drove away. As we left, I saw that a number of locals had appeared as if from nowhere and had been watching our performance. What, I wondered, would they have made of two Englishmen doing their manic, naked dances atop their vehicles. But one thing at least was clear by then – the reason the caves were no longer inhabited.

The survey was finished by the early spring and in the six months we had built up an understanding of the surface geology of most of the province of Tripolitania. The weather was getting hot although, surprisingly, we had experienced some heavy rains over the winter – downpours which made long-dry wadis flow again, and at some of our campsites turned the floors of our tents into footbaths.

In the War the opposing armies had fought over a lot of this country and some parts – the Wadi Zemzem for example – were notorious for the number of unexploded mines remaining. Sometimes of an evening

Laurie and I would wonder aloud whether the armour-plating beneath us would save us if we hit a mine, and the danger was never far from our minds as we went about our daily work. Our peace of mind was not helped by news that Don Sheridan, the geologist running the Kufra survey in the southeast of the country, had just had the two front wheels of his Bedford truck blown off by an old wartime mine.

But we were fortunate and were able to bring the camp back to Tripoli, intact. After two months in the office, writing the survey report and drawing the copious maps and diagrams that went with it, I was to return to England. In May of that year, 1960, Anne and I packed our few household belongings and left our flat on Via Bianchini. With baby Helen we boarded the BOAC plane for Heathrow. The flight had come from Ghana and by chance we found ourselves sharing the First Class compartment with Kwame Nkrumah who was visiting London for the Commonwealth Prime Ministers' Conference. Among his entourage was a big lady in a brightly coloured dress. I remember her particularly, because although she had been allocated a seat beside us, she preferred to spend the entire journey stretched out by our feet, asleep on the floor.

I had gone to Libya a 21 year-old virgin, and now, two-and-a-half years later, here I was with a family and with a fund of traveller's tales to tell. Professionally I was maturing too. It is said that a geologist is only as good as the amount of geology he has seen. By that yardstick I reckoned I had developed a lot on that first overseas posting. I had found ways to work with persons of different temperaments, to survive under some pretty harsh conditions, and to get by in a foreign language. My career was going well, I thought, and most important, it was fun. But looking back I can see how incomplete my experience was. I may have been growing as a geologist but surprisingly, I now see, I remained completely ignorant of the commercial context of the work I'd been doing. I could not have said what the price of a barrel of oil was, how much it was costing to drill an exploration well, or how any oil that BP might discover would be brought to the market. In short, the entire commercial side of the business was a closed book. It would not be for many years that I would be able convincingly to call myself an oilman.

As an aside, I should also record that all of the Libyan concessions which had been awarded to BP in that first round turned out to be largely devoid of oil. The thinking on which those early applications had been made was reasonable at the time, but it was erroneous. From the limited geological data then available, BP's management in London had correctly worked out where the main structural uplifts were located, and since oil tends to migrate upwards, these structurally high areas were thought to be prime areas for oil to accumulate. But as drilling got underway in the 'fifties and the first oilfields were discovered by our competitors, it became clear that our concessions were in areas which had been uplifted too much; the most favourable areas were turning out to be those where the sedimentary rocks were less attenuated, in other words, in the more basinal parts of the country. Acting on this realisation, BP successfully acquired concessions in the Sirte Basin where later drilling brought to light several giant oilfields.

Chapter Five

After more than two years abroad I was due a couple of months' leave, (my nuptial trip having been classed not as leave but as a field-break). We stayed with my parents at Harold Wood and by this time any resentment on my father's part over my injudicious marriage was forgotten. I was treated as a returning hero and my parents doted on their new grand-daughter.

My dysentery was treated by specialists in Harley Street where I was sent by the Company's medical department, and by the summer I was fully fit again and keen to get back to work. BP at that time was working with British Gas to find suitable underground sites onshore in which gas could be stored. It had long been known from surface geological mapping that the Chalk in the region of Winchester formed a large anticlinal fold, several miles long. Buried beneath the Chalk is a layer of clay called the Gault Clay and below that is the Lower Greensand which, it was thought, might be a suitable reservoir rock into which BG could inject its gas.

A pattern of bore-holes had been planned to confirm the suitability of this scheme and I was to be the resident geologist attached to the drilling rig. Anne and I rented part of a large house on the edge of the small Hampshire town of New Alresford. Helen was now a toddler and after the dust of Tripoli and the confinement of our flat there, the sweeping lawns of Sun Hill were a delightful contrast. When there was a lull in drilling and I could get away from my laboratory caravan I would wander the hedgerows with the landlady's shot-gun under my arm, sometimes returning with a rabbit for Anne to stew.

From the Winchester project I was assigned to Kimmeridge. The Jurassic shales which outcrop along this part of the Dorset coast are a rich hunting ground for geologists keen on fossil ammonites, and the strata could be clearly seen to have been folded into an anticline (by the same forces as had produced the distant Alps). Oil had been found a year or so earlier in the first well on the Kimmeridge Anticline and I was to be geologist on the second well, a step-out along the axis of the anticline.

The rig was erected in a field close to the edge of the cliff, near to where a track leads down to the beach. The boundary of a military firing range was a short distance away and drilling was frequently interrupted by their activities, but the well crept deeper and at a depth of just a couple of thousand feet we struck oil. Since, in a sense, this is what I was being paid to look for, it was not before time that I should see crude oil at close quarters. The limestone core that had just been pulled from the bottom of the borehole oozed brown-black stickiness, but it was its smell that I remember best – a slightly sulphurous, earthy, and pleasant smell.

In the summer there was a constant trickle of holidaymakers and I enjoyed explaining to the more inquisitive of them that beneath their feet was indeed a small oilfield. It was too small to justify building a pipeline, but the oil was collected in a large tank which was regularly emptied and taken to Esso's Fawley refinery by road tanker. One of the puzzles of the British oil industry is that the Kimmeridge field has continued producing oil into the 21st Century, yielding in total much more oil than geologists and engineers calculated could have been present in the small Kimmeridge Anticline.

The main focus of BP's onshore effort remained the East Midlands, and for a while I worked on boreholes being drilled at Gainsborough in Lincolnshire. Oil and gas had been found directly beneath the town, and the further wells necessary to determine the extent of the field and to maximise the production rate were being drilled on almost any patch of open ground. Another project was an exploration borehole at Glinton, near Peterborough. This had no real expectation of finding oil or gas but was to obtain geological information. What we found was that the favourable Carboniferous rocks of the East Midlands had thinned to zero at Glinton – an important geological data point to guide further exploration.

This spell of work onshore the UK spanned a period of some eighteen months and during that time I decided we needed a home of our own. Staying in rented accommodation or with my parents was fine up to a point, but these were not long-term answers to our needs. We settled on a modest semi-detached house on a new estate in Billericay. At some time in the future I expected to spend periods working at Head Office, and the rail link to the City could not be bettered. By taking out a mortgage we felt the £4000 purchase price was within our reach. We were immensely proud of our suburban 'semi' and I worked hard in the small garden, laying out paths and flower-beds and planting trees. And looking westwards down the cul-de-sac I could see in the distance the wooded hills around Warley where, as a teenager, I had spent many a summer's evening treacling for moths.

In the early 'sixties that I am writing about, oil and gas exploration worldwide was still largely an on-land business. But it was not difficult to imagine that sedimentary basins extended beyond the shoreline, and with this in mind BP had begun in the 'fifties to look at some of the shallow seas of the world. An obviously attractive place was the Gulf, and negotiations with the government of Abu Dhabi led to the award of a large offshore tract to a joint venture named Abu Dhabi Marine Areas Limited, or ADMA for short, in which BP was the principal shareholder and the operator.

There is inevitably not much preliminary geological survey work that is feasible in an offshore venture, and the limited amount that had been done involved the celebrity figure, Jacques Cousteau, collecting underwater rock samples from a shark-proof cage. Seismic surveys had followed, and when the drilling programme began, the first boreholes confirmed this to be a rich oil province. The first oilfield to be developed was named Umm Shaif, and I learned in autumn 1960 that I was being sent there as resident geologist on Umm Shaif No. 4, the fourth of what would eventually be over two hundred wells on the field.

There were not many of BP's worldwide operations where families were unable to accompany their husbands, but ADMA was one. I flew to Bahrain and stayed overnight in the company guest-house, taking the opportunity before turning in for the night to explore the *souq* and buy a couple of the copper and brass, beaked coffee pots which are a trademark of Arabia. The next day I boarded a light plane and was flown along with a couple of mailbags and various engineering spare parts to ADMA's base on Das Island, some hundred kilometres off the Trucial Coast.

Before ADMA took it over, Das had been a truly desert island: a mile-long, barren, rocky wilderness where fishermen and pearl-fishers from the mainland would occasionally anchor their *dhows* while they searched the sandy beaches for turtles' eggs. The plane touched down on the newly bulldozed airstrip and I could see a harbour had been created at one end of the island, while at the other end was a growing town of workshops, stores, offices and accommodation for the hundreds of men working on the project.

I was allocated a bungalow which I was to share with one of the drillers and with Dickie Burt, the petroleum engineer. It was air-conditioned, comfortable, and only a short walk to the office and to what became a favourite haunt of mine – a cove of white coral-sand and limpid blue water protected from the Gulf's sharks by a rock breakwater. Actually, it was a short walk to everywhere else on the island too.

Das Island was the base, but the drilling-rig was where the real action was taking place. The "*ADMA Enterprise*" was a rig of the jack-up type: a square barge loaded with equipment and capable of jacking itself up from the sea-bed until it was clear of the waves. A helicopter flew daily between the island and the rig and I took the first opportunity to get to work. Dickie and I shared a laboratory on the *Enterprise* and after stowing my gear in my sleeping quarters I got him to show me the ropes. On a land rig it's possible to get away from the worst of the din, but here the clangour was everywhere. Pumps, diesel engines, fans, electric motors, generators – everywhere you looked some piece of machinery had been shoe-horned in, and every machine was contributing to the background hum. And if you closed your eyes and stood still, you could detect the slight swaying motion of the rig itself as it groaned on its four huge legs.

On the one patch of clear deck-space there was a track-mounted crane which itself added to the noise as it unloaded barges bringing materials and equipment out from Das. This had been the site of a particularly ghastly accident a month or so before I arrived. The then petroleum engineer, an Englishman called Howard Rudge, had been leaning over the rail beside the crane as he watched it unloading a vessel moored some seventy feet below. As his eyes were on the barge he was unaware that the crane was slewing around to drop its load on the deck behind him. The massive iron counter-balance which forms the rear of the crane swung around behind Howard and before he could jump clear he was sheared almost in half over the rail. For a while after the tragedy a safety barrier was put around the crane whenever it was working, but it was found to be inconvenient and so after a few weeks was discarded – a lackadaisical approach to health and safety which would be unthinkable today. I need hardly say that even without the safety barrier I kept a respectful distance from the crane whenever I leant over the rail to watch the barges come and go.

Life on the island was made as comfortable as possible for ADMA's employees but it was not a place I liked to linger. More sociable types than I would prop up the clubhouse bar; keen golfers could play on a dusty course which wound between buildings and storage yards and had greens of diesel-soaked sand; there were a couple of tennis courts which were pleasant to play on once the sun's power had eased; and there was the beach and swimming cove. I might spend a few days on the island when drilling operations allowed, but for the rest of the time I was happy to live on the *Enterprise* and put up with the din and the ever-present smell of diesel.

The routine method of travel between the island and the rig was by helicopter, but if you were in no hurry there was always a work-boat willing to take you. A small disadvantage of the boat trip, however, was the matter of getting from the boat up onto the deck of the rig once you arrived. You had two choices: there were rope scrambling nets which were fine if you had a commando background and could face a climb equivalent to going up the side of a seven-storey building; or you could use the personnel carrier. The personnel carrier was a cone of rope netting atop a five-feet wide rubber ring, slung from the crane's hook. It was considered sissy to climb inside the rope cone to be hoisted on board, and so one stood on the ring and gripped the outside of the rope cone with white knuckles as it lifted its passengers high into the sky – perhaps a hundred feet to clear the various machines, hoppers and cabins on the rig – before allowing them to step onto the rig's deck.

The helicopters were large and capable of carrying a dozen or so passengers. On one occasion, I remember, I was the only person waiting to climb on board for the twenty minute journey to the rig. Before I boarded, the pilot came up to me and said that as he had such a light load that day he thought he might practice his emergency landing procedure. Bearing in mind I'd never flown in a chopper before coming to the Gulf, I was a bit anxious to know what this was going to involve. "There's not much to it" he explained, "I'll take her up to a couple of thousand feet and then cut the engine. You'll notice it as we suddenly drop, and the sea will seem to be rushing up towards us, but you needn't worry!" Needn't worry, indeed; I was worried stiff! Wondering whether it might not be prudent to take a later chopper that day, I sought reassurance that I wasn't shortly to become a casualty statistic: "I suppose you've done this emergency drill dozens of times before, haven't you?" I suggested querulously. "Actually, I haven't," he replied with a grin, "but I've read a lot about it."

The Umm Shaif No. 4 well ground towards its two-and-a-half-miles-deep objective, but not without difficulty. For long stretches we were drilling blind, which meant that the mud which was pumped down the drill-pipe (and was intended to hold back any sudden pressure surge as well as to carry the rock cuttings back to the surface) was failing to circulate back up the hole. This in turn meant that I had no samples of rock on which to base my predictions of when we were likely to drill into the high-pressure oil reservoir layer – a potentially dangerous situation. It was clear that fissures and cavities deep below sea-level were swallowing up our drilling mud. The drillers tried to block off these deep mud-losses by pushing all manner of things down the hole, from bales of straw to piles of old rubber boots, but without success. But meanwhile, working in my laboratory with the geological logs from the previous three Umm Shaif wells, I made a discovery. I found that there was a correlation between the electrical logs (vertical profiles of the electrical characteristics and the natural radioactivity of the rock layers through which the wells were drilled) and the logs from the same wells which showed the rate at which the drilling bit penetrated the different rock layers. In this way I was able to use the penetration rate of our present well as a fairly accurate indication of the type of rock we were drilling through, and so predict the depth at which steel casing should be run to stabilise the well before entering the oil zone. Surprisingly, this had not been done before, at least not in the Umm Shaif Field. It was a small technical triumph that I didn't hesitate to report to London.

It was about this time that we received a pair of VIPs, keen to look over these novel offshore operations. Along with a diplomat described as the UK Political Agent from Kuwait (equivalent, I presumed, to an Ambassador) there was a British politician in his mid-forties who was introduced as Lord Privy Seal. (Only later did I find that his particular responsibilities were for foreign affairs, including negotiations which were to lead Britain into the Common Market). Our more senior personnel must have been doing more important tasks, for I was given the job of escorting our visitors over the *ADMA Enterprise*. They were a pleasant, if undemonstrative, pair. We started the tour on the derrick floor, the driving force of the rig since it is here you can watch the awesome power of the giant diesel engines being transmitted to the drill-pipe which, in turn, rotates the drill-bit thousands of feet below. We looked at the mud pits where the rock cuttings are separated from the circulating drilling-mud, and we finished in the laboratory where I explained the geology which underpins every oil exploration project. One of the important strata the well had penetrated was a thick layer of anhydrite, a rock akin to alabaster which, because of its very low permeability, forms the cap-rock or seal above the Umm Shaif oil accumulation. Solid cores of this anhydrite had been taken to ensure we knew precisely its characteristics, and back in the workshops on Das Island the engineers had frivolously turned some of the rock on their lathes to make nick-nacks and souvenirs. I remember presenting an anhydrite ashtray to each of our guests, which they graciously accepted. But I have no delusions that an ashtray was among the treasured *objets d'art* taken by Edward Heath to 10 Downing Street some ten years later to grace his Prime Ministerial desk, for it was he who was the more senior of my guests that day.

Any cessation of drilling meant there was little to occupy the geologist. And so when casing was being run and cemented in the hole, or when the drillers were struggling to regain circulation of the drilling-mud, I could either return to the island or find some recreation offshore. Often that meant fishing. It was easy to catch grouper weighing ten pounds or more, and they were excellent to eat. We fished for sharks too and on a number of occasions the thick nylon line tied to the rail would twang tight and break as some giant fish made off with the joint of meat we used to bait our five-inch hooks. Only once did we manage to catch a shark and then, as we struggled to lift it clear of the water, again the line broke and the six-footer slid beneath the surface and disappeared.

Heavy, wooden, diesel-powered *dhows* from the mainland fished these waters and a couple of times I organized days out on them. The Arabic spoken in the Gulf is somewhat different from Libyan Arabic, but I enjoyed the days I spent with these Sinbad-like sea-dogs, their nut-brown and gnarled old faces wreathed in good-humoured grins as we struggled to communicate. King mackerel two- or three-feet long were plentiful, and we caught them on lines from bamboo poles jutting from port and starboard. These men knew the waters intimately and a sure sign that we were in a favourable area would be the sudden eruption of the sea as an entire shoal of these handsome fish leapt high into the air, this veritable fountain of fish continuing for maybe as much as a minute. Whether the spectacle was caused by some predator or by our presence I wasn't able to say.

One day in early February I was on the rig when I received a message from the head of the operation to return at once to Das Island. A quick helicopter ride and I was in his office. Harry Cannon was a kindly man, an engineer by training, and he broke the news to me that my sister, Judy, had died suddenly. I could hardly believe it; she was only seventeen years old and I had received a letter from her just the previous week. As the news sank in I made my plans to fly back to England on the first available plane. She was a student at Cripplegate Secretarial College in London and had returned home on the evening of 2nd February with the symptoms of a winter cold. The following day my parents called the local doctor who said she had influenza and advised she stay in bed. On the morning of the 4th my mother went into Judy's room in the normal way and found that she had died in the night; the doctor pronounced that the cause of her death was pneumonia.

I got home in time for Judy's funeral and gave what comfort I could to my grief-stricken parents. It is a paradox that in the English language we have the words *orphan, widow, widower,* but no word for someone who has suffered the even more agonizing loss, the loss of a child.

As a memorial to Judy, and because it throws some light on the well-intentioned but severe attitudes of my father, I reproduce here part of her last letter. She was a dear and thoroughly decent person and a loving sister.

11 The Ridgeway — *Monday, 23rd January*
Harold Wood
Essex

Dear Mick,

Here is one of my few-and-far-between letters. It won't be very literary, but full of news and, of course, the odd touch of humour.

I expect that you will know by now that Grandma has died. It's very sad but she had had all she wanted out of life, and I'm sure that's enough ..... The funeral is on Thursday but I'm not going.

A few weeks ago I thought I'd been jilted because I hadn't heard from Chris for nearly two weeks; but one night he phoned and so I invited him to our College dance. I was so miserable to think he had dumped me without saying so to my face. I couldn't believe it when he phoned, but he'd been trying to get me a lift home from Hendon, where I was going to go to a Golf Club dance with him. Since Daddy wouldn't let me stay in a womens' hostel and Chris couldn't get me a lift I wasn't able to go. Our College dance last Friday was terrifically good, and I like Chris a lot more now. At the hospital that day he broke a glass rod into his hand and was rushed to the operating theatre to have a glass splinter removed from his finger. After numerous injections he wasn't feeling too jolly, but he managed to force himself to enjoy the dance. He's awfully clever: he's got three A levels and 8 O levels at G.C.E, ...

He is thinking of getting a scooter, which I think would be marvellous. I happened to mention it at home, and straight away Daddy became all domineering and forbade me to ever go on it. A few weeks ago I posed the question of getting a flat with several other girls, but it was as if asking a brick wall for permission; he refused to discuss it any further, and there's nothing more aggravating than being denied a reasonable argument for something you want to do, to which he is not agreeable. He just thinks I'm getting as much out of him, in respect of College, etc., and then when I'm able to get my own living I want to walk off without any gratitude for his generosity in doing as much as he can for me. He couldn't be more wrong; I just want to be a bit independent. I feel that you must have felt like this at times when you were at home at College. Well, anyway, I want to go abroad after a year's experience. The funny thing is that he'll let me do that!!

Seeing that he was all worked up about me going on a scooter, I thought I may as well put another word in about the flat again. It was even worse than last time. He said that if I mentioned it again he would take me away from College immediately. He thinks College is the cause of my rebellion against home and discipline. As I said before, it's just that I want to experience freedom for a bit. Please write and tell me if you ever felt like that and how you dealt with him in his stubbornness.

Now I've aired my moans I'll proceed to tell the news. On Saturday night Jane and I went to see a Red Triangle play called 'The Matchmakers' at the Lambourne Hall. It was one of those very happy-ending Victorian plays, where everyone ends up marrying everyone else; it was very well acted though. Needless to say this weekend found us very tired after getting to sleep at 2.30 on Friday after the dance, and 1.30 on Saturday after the play.

This evening I drafted a card for the Post Office, saying:

> *A student (girl) aged 17 seeks a Saturday position –*
> *Baby-minding; Shopping; etc. 2/6 an hour.*

I thought I'd try it after getting a bit fed up with working in a shop. I hope I get some replies, as I want to begin soon as I'm desperately trying to save for our holiday – Youth Hostelling in the Lake District. Four of us from College are going – we've just joined. It will be hard going as we are going to walk an average of 8 miles a day. We hope to travel up on a night coach to save expense and a day of our allotted time. This year will be my last long holiday, so I'm going to have as much fun as poss.

Time is running a bit short, and so is paper as you can see from the tattered typing paper I'm using, so please excuse the dog-ears on it.

All my love,

Judy.

Please write to me.

A few days later I returned to Das. The well reached its total depth in spring 1961 and the test programme was carried out. Testing a well means opening the reservoir layer at depth to atmospheric pressure to see whether oil, gas, water, or perhaps nothing at all, will flow to the surface. I felt in some small way that this had become my well, Umm Shaif No.4, and I was pleased that oil flowed at very high rates. Throughout the test, lasting several hours, the crude oil was jettisoned over the side of the rig to form a vast brown slick on the clear blue sea, an appalling act of pollution which was carried out without a second thought but which, like the lax safety measures I have described, would be unthinkable in this day and age.

The Ruler of Abu Dhabi was Sheikh Shakhbut and he took a keen interest in our operations – as well he might as they were shortly to transform Abu Dhabi into a very wealthy country. (Meanwhile it was said he kept his wealth in the form of Maria Theresa gold dollars in chests inside his palace). Among my various reporting duties I was required to send him a weekly report on the well's progress. With the well now completed I wrote the last of my reports to the Ruler, packed my bags, and returned home.

The shattering loss of my sister while I was away on Das Island blighted for many years all recollections of my time there. And for as many years I could not bear the pain of mentioning her death, even to those closest to me. Now, with the emotional scars healed, I still look back with few positive memories of that period of my career. Perhaps it was the sheer artificiality of the working and living environment, the feeling of being hemmed in by buildings, machinery and other man-made paraphernalia. Among the few joyful times I remember were when I was snorkelling in the crystal-clear waters above the fish-teeming coral reefs, or my days with those simple Arab fishermen.

Chapter Six

In the UK again I was put back to work on various drilling projects in the East Midlands and in southern England, returning whenever I could to Anne and Helen in our new family home at Billericay. As I was beginning to wonder where I might next be posted, Head Office let me know that I was to be sent to New Zealand. This was the best news I could have received.

The usual liaison with Personnel Department got under way on such things as salary, overseas allowance, housing arrangements and travel. Anne was by now expecting our second child and for a while this posed a problem; airlines had rules governing the time during a woman's pregnancy when she may not travel, and if BP wanted me to be in New Zealand by the end of the year it looked as though Anne would have to stay behind. But being 1961, shipping companies still ran regular passenger ships to the dominions and it was agreed that in view of the unusual circumstances I and the family could travel to New Zealand by liner. What a thoroughly decent company BP is, I thought, and I've thought so ever since.

Before the tickets could be issued there was inevitable paperwork to be completed for the shipping company, and I remember that one of the forms required passengers to state their full name and any honours they may have. Feeling quite a man of the world by now, I duly completed the documents and posted them back to the company. I still shudder with embarrassment to recall the passenger list which was displayed throughout the ship and issued to every passenger – among the knights, OBEs, Lords and Ladies there was apparently just one passenger with an honours degree: Michael Frederick Ridd, BSc.

On the scheduled departure day in November 1961 my parents drove us to Tilbury where P&O's liner *"Orsova"* was tied up at the dock, and embarkation was underway. After emotional farewells (for we didn't expect to see each other again for another couple of years) the three of us climbed the First Class gangway and were shown to the cabin which would become our small world for the next six weeks.

No one could pretend she was a beautiful ocean liner, and her unusual witch's-hat funnel added nothing to *Orsova's* charms. But on board she was comfortable and our cabin, Number A1, was as good as they got. No departure by plane or train can compare with the drama of a liner slipping her moorings and inching away from the jetty. We, of course, were among the crowds of passengers lining the rail, and the side of the ship was festooned with paper streamers linking those onboard with the friends and relatives on the quayside. As the gap widened the paper streamers would snap and the evening was loud with tearful shouts of goodbye.

As we headed south toward our first port of call we fell into the ship's routine – promenade on deck after breakfast; visit the ship's library; join in the daily lottery to guess the mileage run the previous day; change for dinner (black tie and dinner jacket, of course, except on the first night out from port). We had a day ashore at Gibraltar, but on the next leg of the voyage *Orsova* was hit by a severe Mediterranean storm. For twenty-four hours she was tossed like a cork, and we were too sick to leave our cabin. Lying on the hard cabin floor was safer and less uncomfortable than lying in bed, but at times the vessel rolled so violently that all three of us would slide across the floor to the wall, and as she rolled again we would slide back to the other wall.

It was a relief to get ashore for a day at Naples where a tour to Pompeii had been arranged. Then to Port Said, and through the Suez Canal to Aden. My memories of the Indian Ocean are of lazing by the pool or watching flying fish leap from the water by our bow, guessing for how long they could glide before dropping again into the sea.

For me, Colombo was in effect the end of my cruise, because the day we left there I became ill. I developed a fever, my neck swelled and, more painfully, my right testicle swelled to the size of a jaffa orange. The ship's doctor pronounced I had mumps and I was confined to the hospital for the next couple of weeks. I remember little about it now except that the pain required that I had a daily shot of morphine. The medical term for an infected and swollen testicle is *orchitis,* and when I finally felt well enough to think about such things I resolved that, in the unlikely event of ever being asked to name a P&O ship, I would call her the *Orchitis,* to go with their other ships, the *Orsova, Oriana* and *Oronsay.*

We had a stay of several days in Sydney, tied up in the shadow of its famous harbour bridge. Australian newspapers were brought on board and carried headlines of an oil discovery just made at Moonie in Queensland. After a number of earlier false dawns when apparent discoveries failed to lead on to actual production, Moonie turned out to be the first Australian oilfield. Of course, I had no inkling at the time that I would be immersing myself in the Australian oil industry some ten years later. By the time we finally docked at Auckland I was fit again – although I felt weakened for several more months by the bout of mumps. My right testicle had shrunken to half its normal size and has remained wizened ever since. We boarded a flight on a Douglas DC3 and were met at Gisborne airfield by BP's Senior Geologist, Peter Phizackerly.

Gisborne was a sleepy town of some twenty thousand souls on the eastern side of North Island. It's not on the way to anywhere important, and in the early 'sixties was frequently cut off from the larger centres of Hawkes Bay to the south and the Bay of Plenty to the west, when landslides in the Wharerata[2] Hills or in the Wioweka Gorge closed the roads. The area was almost totally dependent on agriculture, and in particular sheep farming.

On the face of it, Gisborne would appear a surprising place to base BP's oil exploration campaign. The Company (the appointed operator for a joint-venture comprising Shell and a New Zealand company called Todd) had been awarded concessions over the entire east side of both the North and South Islands, and could have set up its headquarters at any one of a number of larger and more accessible towns. The reason for choosing Gisborne was that natural seepages of gas and crude oil occur at several locations in the countryside around Gisborne and it was reasonable to focus exploration work on that area. After all, the essence of petroleum exploration is to lessen the inevitable risk that when a well is drilled it will fail to find oil or gas. That risk is an amalgam of many unknown factors:

1 Was oil or gas ever generated in the sediments that make up this basin?
2 Are rocks present with suitable porosity and permeability to constitute a reservoir in which the petroleum can accumulate?
3 If petroleum did form, did it migrate out of the so-called source rock into the reservoir rock?
4 Is there a trap which will allow the migrating petroleum to accumulate instead of leaking away to the surface?
5 And did the hoped-for petroleum migration take place at the appropriate time with respect to the time at which the hoped-for trap developed?

[2] 'Wh' in Maori place-names is generally pronounced as 'f'.

Given that catalogue of risk factors it can be seen that oil or gas fields are virtually a fluke of nature, and it is not surprising that the majority of exploration wells are dry. Therefore for BP to have focussed its search on an area where it could be certain that at least the first of the above requirements would be met made good sense.

The Administration Manager at the Gisborne office had rented a furnished house for us on Russell Street and we settled in without delay. It was a painted wooden house with a roof of corrugated iron, and it stood in a large lawned garden. The landlord had thoughtfully left me his lawnmower – and not simply a push mower as I was used to in the UK, but a petrol-engined one which turned out to be the norm in New Zealand. Being the height of summer the giant magnolia tree in our garden was in bloom, its creamy-white flowers as big as cabbages. And of an evening as dusk fell we could hear the shrill call of the *wekas* from the bush beyond the garden, near-flightless ground-living birds not unlike the better-known kiwi. We were delighted with our good fortune at having been posted to such a green and pleasant town.

We had not been in our new home more than a few weeks when Anne gave birth to our second child, a daughter who we named Sally Clare. She was born in the Lister Hospital on 27th January 1962. Even in New Zealand where midwifery services were second to none and enlightened attitudes generally prevailed, it was still not the custom for fathers to attend the birth of their children, which over the years I have come to regret.

The office was a large converted house on the north side of town and I was allocated a room in what had been the garage. The staff of about a dozen were a charming bunch of people. The top man, General Manager Roy Greenham, was an easy-going Australian who had earned his stripes working in Papua New Guinea as a geologist. Peter, the Senior Geologist, had recently joined BP from one of its subdiaries in Trinidad and he regaled us with his geological experiences there whenever we tried to discuss with him the geology of New Zealand. He was a bachelor, and with his good looks and urbane manner he must have been one of the most eligible men in the whole Poverty Bay District. The small team of geologists included two I had worked with in Libya, and I was pleased to join up with them again: Jim Hawkins and my good friend Ron Walters.

It was Peter's responsibility to allocate the geological projects, and mine was to be an area lying to the north of Gisborne, from the Maori coastal settlement of Whangara extending inland to include the drainage basin of the Waimata River. I was issued with a Landrover and a young Gisborne man named Bruce was to be my field assistant. I could have had no-one better than Bruce – he had the build of an "All-black" and a willingness to scramble up any hillside, wade any river, and then at the end of the day carry a heavy rucksack full of rock samples back to the Landrover.

For the first week of the Whangara-Waimata survey we quartered the area, trying to understand the essentials of the geology. It was a depressing week – over parts of the area the rocks outcropped as a chaotic jumble of sticky mud and boulders, and in the other parts, where the rocks were stratified and seemingly orderly, they appeared all to be the same endless grey mudstones and sandstones. I began to despair that I would ever be able to unravel its structural complexity or make sense of the apparent uniformity of its boring grey beds.

A typical day would see Bruce and me heading out of Gisborne by 8.30 a.m., parking the Landrover at a sheep station or as close to the planned traverse as we could get, and then for the rest of the day walking the hills and streams while I made observations of every rock outcrop we came across. It was physically demanding work, particularly at the height of summer, but the farmers we came across (or the 'graziers' as they were known in New Zealand) were friendly and hospitable folk, and many were the times we would be invited to call at the homestead for refreshments. In those days Britain was still touchingly referred to as "Home" and we were greeted enthusiastically, the housewife invariably bustling around and saying "Now let me see what I've got in my tins," which meant checking to see whether any of the cakes she'd cooked last week were left, so that we could be offered proper hospitality along with a cup of tea.

The hills of eastern New Zealand were forested when the *pakeha,* the Europeans, started settling there in the 1800s, but, except in the more remote parts, that tree cover was progressively burned off to provide sheep grazing. Charred skeletons of *rimu, totara* and other tall native trees still dotted the landscape as dismal reminders of that early vandalism, while a lasting legacy is the region's vulnerability to landslides now that the tree cover has been removed.

Generally we would plan our daily traverses along the many stream courses which dissect this steep hill country. They run in deep valleys, and in the headwaters there are many waterfalls which had to be negotiated. It is in these streams that the least-weathered outcrops occur, and the rocks' nature can therefore more readily be seen. Gradually, as the months went by I began to make some sense of the geology. This process was partly a matter of "getting my eye in" but it was enormously helped by having aerial-photograph coverage of the area and, even more, by the work carried out in our Gisborne palaeontological laboratory by Ron

Walters. The rocks of my survey area were predominantly of Tertiary age, and had been laid down on the ancient sea bed. By extracting the microfossils they contained, and applying principles of evolution, Ron was able to give me a fairly accurate indication of the relative ages of the different rock samples I collected. Where a succession of strata has been deformed by faulting and folding, this ability to say which rocks were originally at the base of the pile (that is, the oldest) and which were at the top (the youngest) is essential to elucidating their structure.

Summer gave way to autumn and the weather turned wet. The streams and rivers rose, and each evening we would return home soaked to the knees from another day's wading. The grassy hillsides which we had been able to drive over without a second thought became slippery and hazardous, and Bruce and I became adept at winching the Landrover out of trouble.

But we did have one incident – I recall it vividly – when a river crossing got the better of us. We were working on the coast north of the remote Maori settlement of Whangara. With the tide out we drove for a mile or so along the firm sandy shore which curved beyond the settlement, until we came to the mouth of the Pakarae River. This separated us from the rocky foreshore where we planned to carry out the day's traverse. We nosed the Landrover into the river and, finding that the water rose no higher than the wheel hubs, we inched our way to the other side. After parking the vehicle well clear of the water on the north side of the river mouth, we set off northward toward Gable End Foreland and spent a successful day, since the rock outcrops in the sea cliffs and on the foreshore were excellent. By late afternoon when we returned to the Landrover the tide had turned, but the river mouth was still shallow enough to ford. Again, we inched forward into the river and had no trouble until we got about halfway across. We must have encountered a patch of soft sand, because the vehicle lost traction and the wheels started to spin. The technique in that situation is to take your foot off the accelerator immediately to avoid digging the vehicle deeper into the sand. With the engine still running we got out and paddled round the Landrover to see what could be done.

Across the river on the south side we spotted a large tree trunk which had been washed up to the high-tide mark and this, we thought, would provide the anchor we needed to winch the Landrover out of trouble. It took just a couple of minutes to hook up to the tree trunk. Co-ordinating the pull of the winch with the power to the wheels, Bruce let in the clutch, while I stood in the river directing operations. The winch cable twanged tight and the Landrover moved a few inches, but then, instead of moving further, we saw that the tree trunk was being pulled across the beach toward the river and the Landrover remained firmly stuck.

This was now beginning to look serious. The tide was coming in and with the winch we had played our last card. Then I thought of the Maori settlement a mile back along the beach. Leaving Bruce with the Landrover I set off at a run. They were a friendly bunch: big, jolly, laid-back men who offered to help without a moment's hesitation. A couple of horses were rounded up and we headed back toward the river mouth. But even with the winch, the wheels slowly turning and the two horses pulling with all their might, we still failed to shift it.

By now the water level was upto the Landrover's floor and it would not be long before the engine was flooded. The only chance of getting her free now was to return to the Maori settlement and phone the office in Gisborne for help. I ran off again along the beach, made the phone call and waited for help to arrive. Forty minutes later a second Landrover bumped up the track to the Maori settlement and I climbed in gratefully. The tide was now too high to drive north along the beach and it took some time to work our way through the sand dunes. Then, as we rounded the last grassy dune, the sight that confronted us made my heart sink. The tide had come in and what was previously the river mouth was now a wide inlet of the sea. And out in the inlet all that I could see of my Landrover was its roof, lifting and swaying as it was caught by the incoming waves. It was dark by the time I had swum out to connect the winch cables of the two vehicles and pulling could begin. No doubt the fact that my Landrover was now buoyed up by the sea made our job easier, and steadily she was tugged free of the waves, cascades of seawater pouring from the open windows.

The next day I steeled myself and knocked on the boss's door to report the night's events. Since I expected the Landrover's engine and all its other moving parts would be a write-off, I fully expected a thorough dressing down. After hearing me out, Roy's face broke into a grin. "No worries, Mike" he said, "these little accidents do happen". And far from berating me he went on to describe an adventure he had had in Papua New Guinea when he had flooded not a Landrover, but a Catalina amphibious aircraft. What a thoroughly decent bloke, I thought, as I left his office and returned to my cubby-hole room out in the garage.

The geological interpretation of my survey area which gradually took shape on the map and in my mind might best be explained by imagining a huge meat pie, a pie some twenty miles wide having a pastry crust many thousands of feet thick. Imagine then the weight of the crust causing it to sag in places, allowing the meat and gravy to push their way to the surface. In the case of the area I had surveyed, the material which pushed its way to the surface was a mass of plastic clay and the process had been accompanied (and perhaps facilitated by) methane gas which continues to the present day escaping as natural seepages. I found from historical records that earthquakes had sometimes triggered violent eruptions at the sites of some of these seepages, when onlookers had observed mud and boulders being ejected scores of feet into the air – doubtless an impressive spectacle. In reality, the geological picture which emerged was rather more complicated than I've described. Crossing this so-called "pie" are a number of major faults, as if the entire pie (including the pie-dish) had been cut through and the separate parts had been moved relative toward each other.

Back at the office in Gisborne I wrote my report on the survey, the drawing office drafted the maps and diagrams, and the package was sent off to London. Although I had to conclude that I could see no realistic chance of finding commercial quantities of oil or gas in the area, the Chief Geologist, Norman Falcon, found my geological analysis interesting. He suggested, furthermore, that I should consider publishing it in the form of a technical paper. To receive these comments on my work from someone I respected and admired as much as Falcon was immensely satisfying. I duly re-wrote it in the prescribed form, submitted it to the editor of the New Zealand Journal of Geology and Geophysics, and in due course it appeared in print. I was 26 years old and that was a milestone in my career. To have unravelled the geology of a very complex area (whether or not my interpretation would prove over time to be correct), and then to see it published, gave a massive boost to my professional confidence.

Much though we enjoyed living in the house on leafy Russell Street, (and later, for a while on Domett Street) the opportunity to move to Waikenae Beach came up, and we grasped it. Separated from the surf by just a few metres of marram-covered sand dunes, the house had two storeys. Its glass-fronted upper floor was like a ship's bridge, with views across the breadth of Poverty Bay to Young Nick's Head. It was on this very beach that Captain Cook first set foot on New Zealand soil on 8th October 1769. After skirmishing with the local Maori he withdrew without obtaining the food or fresh water he was seeking, and so he gave the bay the name by which it's still known. It was a splendid place to live and we saw out our time there, hosting swimming parties and barbecues in the summer months and being lulled to sleep at night by the pounding of the surf.

New Zealanders are enthusiastic yachtsmen and soon I was infected with a wish to build myself a sailing dinghy. Conversations with my Russell Street neighbour, himself a leading light in the local yacht club, led me to decide on a Z Class dinghy which he thought was a class likely to increase in popularity. I bought the plans and set about building it in the garage. The ideal timber would have been *kauri* but this fine and durable wood was now virtually unobtainable, and so instead I settled for *kahikatea*, another native wood which is even-grained and easily worked, although lacking the durability of *kauri*. Most of the materials I needed could easily be bought at timber merchants and ironmongers in Gisborne, but I remember on more than one occasion being told that the waterproof glue, or the bronze screws I asked for, "aren't in at present, but we're expecting a shipment from Home in a month or so." It took nine months before she was finished and we christened her "*Imshi,*" Arabic for "Go!" and launched her on Gisborne harbour. Ron and I sailed her on Poverty Bay in all weathers and, although we never had much success at racing, she was a safe and seaworthy little vessel. But the hoped-for explosion in popularity which had been predicted never did materialize, and so there were never more than a handful of Z Class yachts to be seen on the bay.

The launching place was close to Kaiti Beach, on the east side of the bay and close to the meat works. Truckloads of sheep would arrive here from the surrounding countryside, and refrigerated ships – "Homeboats" the local people called them – would anchor offshore to await their cargoes of New Zealand lamb. The effluent from the meat works was piped directly into the sea, which meant that when processing was at its peak a noticeable meaty smell would waft over Gisborne, and a red tide would spread across half the bay. I presume that this practice ceased years ago, but in the early 'sixties it was given little thought and was accepted by the local people as a fact of life.

Apart from sailing together, Ron and I would enjoy knock-about games of golf, often rising at dawn to get in a few holes before the office opened. He was a far better player than I (not just at golf, but at most games) and I remember how he would savour the pleasure of a good shot onto a still-dew-covered green. At weekends our two families would meet for a meal at his house or ours. I wish I were able to recall that we enjoyed many a bottle of good New Zealand wine together on these occasions, but in truth it was poor stuff, a fortified faux-Madeira from the Waiherere Wine Company being all that was available. While the ladies talked, Ron and I would repair to an adjoining room where he would take out his Penguin book of symphonies and, analysing each movement, we would work our way through the latest classical recording borrowed from the Gisborne Library. Ron opened my eyes to many pleasures of the mind and I still mourn his premature death some twenty years later.

With the Whangara-Waimata geological survey completed, Peter asked me next to carry out a survey of South Canterbury. It was a much larger area than my first survey and extended around the rim of the Canterbury Plain from Pareora in the south to the Rakaia River at about the latitude of Christchurch. Bruce and I set off by Landrover (now refurbished after its immersion) and after crossing Cook Strait by ferry, and driving south through Marlborough, we arrived a couple of days later in Christchurch where we established ourselves in a fine Edwardian hotel, the Clarendon.

The Canterbury Plain itself is flat farmland crossed by broad, braided, rivers which drain the Southern Alps. There was little scope for geological fieldwork on the plain itself for the simple reason that the bedrock is concealed beneath the alluvium. But in the foothills of the Southern Alps which form a rim around the west side of the plain, there are small basins of Tertiary rocks and it was these that we had come to study. The purpose of our survey was to allow the company's geophysicists to interpret the seismic data that they were about to obtain from their seismic survey over the plain.

Those foothills must be some of the most scenically beautiful parts of New Zealand. It was 1963, and in the months we were there the tussock-grassland hills were gold in the winter sunshine; stands of pine or macrocarpa sheltered sheep stations in the valleys, and the backdrop was the snow-covered Southern Alps.

The work had none of the technical challenge of the geology around Gisborne. There was no point at which the scales fell from my eyes as happened when I unravelled the complexity of the Whangara-Waimata area. But it was enjoyable nonetheless. We stayed at small country hotels in the towns which dot the region – Mount Somers, Fairlie, Geraldine, Albury, are some that I recall – hill-billy towns looking as if they had been erected as film-sets for some western movie. The accommodation was plain but the food was just what hungry workers needed after a day walking up ice-cold streams. And one fine British tradition which these pubs maintained was to wake their guests with a tray of tea each morning, accompanied not by a biscuit but a slice of bread and butter.

New Zealand's licencing laws required pubs to cease serving alcohol at 6.0 p.m. but if the party was going with a swing it was not unknown for the publican to close the front door and pull down the blinds. Drinking might then carry on till all hours with little chance of being interrupted by the district policeman – a certain recipe for a headache the next day.

I'd been asked to complete the survey in about two months and so we worked long hours, often trudging back down some stream to the Landrover after nightfall. The bright winter sunshine would give way to a cold and starry night, and frost was quick to coat the ground. Bruce and I were used to returning to Gisborne wet to the knees or worse, but soaked trousers in Canterbury meant that as the temperature dropped, our legs would be encased in stiff, leathery, tubes of ice.

Working higher in the foothills than usual one day we ran across the ski club of a local small town. The snow conditions were good and someone was on duty to hire out skis and boots, and so Bruce and I decided on the spur of the moment to take a day off work. The facilities were basic in the extreme – a diesel engine at the foot of the hill driving a continuous length of rope. We were each issued with a gadget like a giant nutcracker which was attached by a belt around our waists, and you got to the top of the ski-slope by grabbing the moving rope with these tongs and allowing yourself to be pulled up the hill. Neither of us had ever set foot on skis in our lives before, but we skiid all that day until we were ready to drop with exhaustion. By evening I had just about mastered snow-plough turns and had become infected with the skiing bug, going on in later years to ski in many parts of the world, albeit with more enthusiasm than grace.

My third and last project in New Zealand was a survey of the central Wairarapa district. The Wairarapa is that stretch of country running southwest from Hawke Bay toward Wellington and my survey was roughly between the towns of Dannevirke in the north, and Masterton in the south, and from the Tararua Range eastwards to the coast. Topographically it is like the Gisborne area, with steep grassy or forested hills and deep valleys with plenty of waterfalls in their upper reaches. But whereas the hills north of Gisborne have a jumbled pattern, the hills of the Wairarapa are aligned in northeast-trending ridges which reflect the structural grain of the rocks, largely imposed by the important northeast-trending faults on which some of New Zealand's earthquakes have taken place in historical times. There were a number of natural seepages of gas, and my task was to work out the geological history of the region and assess whether it had the potential for commercial quantities of oil or gas.

Although there is a liberal scatter of sheep stations over the Wairarapa, there are large tracts which lack any settlements bigger than a hamlet. Added to that, the gravel roads were narrow and wound their way through the hills, which made driving any distance slow and laborious. The answer to our accommodation needs was therefore a caravan. We towed it behind the Landrover, and friendly farmers would allow us to park it beside their shearers' quarters. At the end of a day in the hills, this arrangement allowed us to enjoy the comfort of a shower, an amenity lacking in our small caravan.

Bruce and I had our respective duties when we returned to the caravan of an evening. I would plot the day's geological observations on my maps, label and prepare to send off to Gisborne the rock samples we had collected that day, and ponder the significance of what we had seen. And it was Bruce's job to cook the steaks or the lamb chops – our staple food – which we had bought fresh from the farmer's wife. While Bruce and I generally got on very well together there was one matter which came close to causing us to fall out. After he had cooked the meat he would place it on plates and would unfailingly hand me the smaller of the two portions. I put up with this for a week, and then could stand it no longer – (I'm talking here about two permanently hungry young men). A new procedure was necessary if we were to avoid an ugly scene. And so it was agreed that every night he would serve the food onto the plates, as before, but that I would choose whose was whose. Harmony was restored and it was noticeable from that day how the portions became virtually identical in size.

Everyone who is old enough can recall where he or she was when they heard the news of President Kennedy's death. With the time-difference of 17 hours between Dallas and New Zealand it was the evening of 23rd November 1963 by the time I heard it on the NZBC radio news. Bruce and I were in the caravan, relaxing after a hard day's walking. We were shocked, of course, but there was nothing in our daily routine to be disrupted by the news.

The summer days got hotter and we would steal short breaks for a refreshing dip at one or other of the remote Pacific coast beaches or – a favourite of ours – the river which flows through Masterton where, at the edge of town, we could dive from the overhanging limestone cliff into a deep natural pool. And talking of Masterton, the local radio station there got wind of oil explorers in the area and I was invited to give an interview. It is a natural human tendency to imagine that if exploration is taking place in one's district, it must be because oil is expected to be discovered. Answering questions from a pleasant young woman reporter, I was glad to have the opportunity to put our activities into perspective and quash any speculation that a Texas-style black-gold rush was in the offing.

By early 1964 the survey was finished and Bruce and I returned for the last time to Gisborne. Writing the report took several weeks. Ron's palaeontological findings had to be integrated into the geological picture, and a large number of maps and diagrams had to be drafted and coloured by hand in the drawing office. One purpose of a study of this kind is to reach an understanding of the geological history of the area you have mapped – indeed, the ability to deduce the geological history is a test of whether or not your map is likely to be an accurate portrayal of the rocks that are present, bearing in mind that geological mapping relies heavily on interpreting what *may* lie concealed beneath the surface in those places where no rock outcrops are present.

A pretty satisfactory geological history emerged – periods of marine sedimentation; oscillations of the sea-floor allowing sometimes limestone, sometimes mudstone or sandstone to be laid down; periods of local uplift and erosion which produced breaks in the succession. But one part of my map troubled me. I had plotted the line of a fault (which I named the Pongaroa Fault, from a nearby village) and had noted that on opposing sides of the fault the successions of rocks were dissimilar. I could show convincingly that the fault was a long-established one by the fact that it plunged beneath an unfaulted cover of rocks which themselves were some ten million years old, so the fault was active at least that long ago. One explanation for the differing rock successions on opposing sides, I figured, was that the fault repeatedly changed its sense of movement: first up on the west side then up on the east side, up again on the west side, and so on. But intuitively I felt that this explanation would not do; it was too complicated.

I began to lie awake at night worrying about it. I discussed it with Ron, with Peter, and with whoever else would listen to me, but a simpler and more rational explanation eluded me. This went on for a couple of weeks and I began to think that although it was a complicated and inelegant interpretation, perhaps it was nevertheless the right one.

Then one night the correct interpretation dawned on me. I suddenly saw the light. This must be how Watson and Crick felt when they hit upon the double-helix structure of DNA, I thought. It was an exhilarating feeling and I knew at once it was correct because it was so simple and elegant. The Pongaroa Fault did not yo-yo up and down as I'd thought at first. By stripping off (in my imagination) the cover of younger rocks I could see that the succession on one side of the fault was identical with the succession on the other side – *but about ten kilometres away along the fault* on the other side. In other words, between about twenty million and ten million years ago the Pongaroa Fault had moved with a horizontal wrenching motion, displacing the crust laterally by about ten kilometres. Eureka!

At a number of places around the rim of the Pacific Ocean, recent fault lines have been mapped on which lateral displacements of this kind can be detected whenever there is an earthquake. The San Andreas Fault in California is a well-known example, and in New Zealand the earthquakes in historical times have offset stream courses and even caused dog-legs in roads and fences. And so what my modest discovery showed was that the same lateral fault movements have been taking place on and off for millions of years.

I had moved on from New Zealand by the time I had written-up this interpretation in the form of a technical paper. I sent it to London and, again, was pleased to get Falcon's go-ahead that I may submit it for publication. BP's policy on publications (or Falcon's, at least) was a liberal one – provided that what you wished to publish was of no commercial benefit to possible competitors, and provided that you wrote it in your own time, then there was no objection to you publishing the results of your work.

By then I had been in New Zealand for over two years and it was felt by London (I use the term to include those in Personnel Department and Exploration who determine such things) that I was ready for another move. In any case that phase of exploration was coming to an end. We had drilled several wells without finding oil or gas, and London was losing heart; it was time, they thought, to farm out the concession. Farm-outs are a standard feature of the exploration business: after a company's initial foray, if it has failed to make a discovery it will frequently seek another company to join the venture and carry out further work at its own cost and thereby earn itself an interest. In this case the company that wished to farm-in was the French company, Aquitaine. BP's expatriate staff dispersed, and the office in Gisborne was occupied by Frenchmen.

For me, the posting had been a satisfying and enjoyable one. The outdoors lifestyle and the uncrowded nature of the place were much to my liking. Yes, it had the feel of a backwater, and that might have palled if we had remained in the country for longer, but I knew that when I left I would miss the slow pace of life and the self-sufficiency which kept them driving twenty-year-old cars. And I'd miss the homely energy which ensured the housewives' tins were stocked with home-made cakes. As for my development as a geologist, I had been fortunate to experience a wide range of challenges and I'd found the work more fulfilling than any I had done before. I left for my next posting with my self-confidence much enhanced.

The historical context of that period of exploration is that over the subsequent forty years the entire East Coast basin of New Zealand has remained a disappointment. Companies attracted by the natural seepages of oil and gas have come and gone (including the company, Croft, which I formed in 1985 – of which more anon), and wells have been drilled onshore and offshore, but success has eluded them. There are valid technical reasons why exploration in that basin is particularly risky, but with improving technology I would not rule out a discovery being made some time in the future.

I sold my dinghy "*Imshi*" and our aging Ford Consul, and with some reluctance we left our house on Waikanae Beach. I was wanted in

Alaska. More specifically, I was asked by the manager of the Alaska project, Alwyn Thomas, to come for a short period to take part in a summer survey on the North Slope. In the event my "temporary" secondment turned into a posting which lasted over two years.

Chapter Seven

We broke our journey home at Sydney, and while completing the immigration formalities at the airport an Australian leaned over and announced in a strong Aussie accent "Gooday, I see we share the same surname!" He had presumably noticed the name on our old-style passports. We talked briefly and noted each others' details before flying off for a brief stop-over in Hong Kong. Once back in England I mentioned this coincidence to my father's brother, my uncle Bernard. The name 'Ridd' is an uncommon one and Bernard's hobby was piecing together the family tree. He had been a banker and on retirement had moved to North Devon to be close to the heartland of the Ridds; after years of research his knowledge was encyclopaedic. "Ah, yes" said Bernard, without hesitation, "he will be the great-great-grandson of Samuel Ridd (or whoever it was) who was transported to New South Wales in ..." I have forgotten the details.

Perhaps I should say a little more about the family's history. Although the name is uncommon generally, Ridds are fairly numerous in Southwest England. Various parish churches in North Devon record the name, and in some graveyards it is possible to find the stones of some of our forebears. It is no surprise that when the Victorian novelist R.D.Blackmore wrote his romance of Exmoor he chose the name Ridd for his hero. (Whether that story has any basis in fact I doubt, although my father could not be shifted from the belief that it has). The most interesting branch of the family tree shows that the mother of my great-grandfather, Thomas Ridd, was a Mary Sloley. Her great-great-grandmother, in turn, was a certain Mary Luttrell (born 1682, died 1728).

The Luttrells are one of the blue-blooded families of the region and Dunster Castle was their seat until it was passed on to the National Trust in the late Twentieth Century. Their family tree is well documented and from Mary Luttrell a branch goes back in time to the Paganels, the de Courtenays and the Plantaganets, most notably King Edward I. In view of my later sojourn in Scotland and affection for that country, it is wryly amusing that I should be descended from one whose epitaph in Westminster Abbey reads *Malleus Scotorum*, "The Hammer of the Scots."

On 19th May 1964, after a month or two of accrued home leave, I flew out to Los Angeles, BP's base for the Alaska operation, leaving Anne and the two girls at our home in Essex. The Company put me in the Plush Horse Inn, conveniently located at the foot of the Palos Verdes hills, and as plush an inn as I had ever stayed at. The next day Jim Spence, BP's Chief Geologist for Alaska, picked me up in his large Chevrolet. Jim was an amiable and relaxed Glaswegian with a fund of rude jokes, and I took an immediate liking to him. A few minutes later we were in Palos Verdes Estates, a flower-decked plaza of exaggerated Spanishness and, as the realtors would be telling me a few months later, one of the most sought-after districts of Southern California. BP's offices were on Via Tejon, a stone's throw from the plaza, and Jim showed me around. The contrast with the cubby-hole room I'd occupied in the converted garage in Gisborne could not have been greater. A fountain played in the central courtyard, and our rooms led onto balconies from which bourgainvillea hung in bright profusion. Throughout the suite of offices, background music softly played. It was impossible not to be excited by the sheer luxury of it all.

The weekend after I arrived, Alwyn Thomas gave a party at his home in Palos Verdes and it was a chance to meet the people I'd be working with, along with their families. Alwyn himself was one of BP's most senior geologists although he had moved sideways into management, an inevitable progression for anyone wishing to reach the highest echelons in the Company. Reporting to him was Exploration Manager, John Zehnder, an Australian who had briefly become famous some years earlier while working as a geologist in the Papua New Guinea highlands – the world's press picked up that he had discovered and entered what they dubbed "Erewhon," a misty amphitheatre ringed by high limestone cliffs, whose local tribe until then were unknown to the outside world. Jim Spence was there with his new bride, Sally; and then there were around half a dozen geologists and perhaps half that number of geophysicists. It was a poolside party; the oleanders were in flower; the Californian sun shone; the beer was cold and the food good; and the company was most agreeable. The thought that I was being paid to lead this sybaritic life seemed almost too good to be true.

For the weeks before I was due to fly to Alaska I stayed at the Eden Roc, a serviced apartment block on Redondo Beach. With its pool and its modern American kitchen equipped with gadgets I'd never even seen before, it was as far outside my experience as had been the luxury of the Plush Horse. I would walk the couple of miles to the office of a morning. But walking was not something that the average Californian cared for, and indeed, in residential areas there were generally no "sidewalks." On my way to Palos Verdes, motorists would stop and offer me a lift, assuming, I suppose, that my car must have broken down. When I thanked them and said I was happy to walk, they would look at me with expressions of tolerant puzzlement, no doubt associating my strange behaviour with my English accent.

By mid-June it was reckoned that snow conditions in the Brooks Range would allow geological fieldwork to begin and so the field party flew to Alaska. The departure point for the Sadlerochit Survey was Fairbanks, a town which still retained a flavour of its gold-rush past. The Party Leader was genial Geoff Larminie, a large and colourful Irish geologist with red hair and moustache, the gift of the blarney, and usually seen puffing on a tobacco pipe. Besides Geoff there were five geologists: David Jenkins, Malcolm McClure, myself, and two US geologists on secondment from out partner company. The helicopter pilot and helicopter mechanic had flown on ahead to rendezvous with us at our base-camp; and the remaining member of the team was Hog-Too, the cook.

The night before our departure we hit the town. Most of Fairbanks's nightspots were on Fourth Street and we enjoyed a steak dinner of Alaskan proportions before rounding off the night at the Silver Dollar and the Gold Nugget bars. It was a memorable launch for our survey, although midsummer nightlife in Fairbanks has something of a surreal quality – emerging onto the street at midnight means emerging into broad daylight.

Jim Magoffin's Interior Airways was to provide air support throughout the survey and the next morning Jim was at the airfield to help load the DC3 that would take us and our camping gear to the Brooks Range. He was a laconic bush-pilot with long experience, and nobody could have inspired greater confidence when it came to handling a plane in the Alaskan wilderness. I have seen Jim at the controls of a float-plane taking off from a frozen lake where the lead of open water ahead narrowed to just a few feet; unfazed, he tilted the plane and it continued on one float along the narrowing lead, gaining speed until it could rise into the air as the lead closed beneath him.

Droning north across the Arctic Circle, the rolling forested lands of central Alaska slowly gave way to the bare mountains of the Brooks Range, stretching snow-covered eastwards to the Canadian border and beyond into the Yukon Territory. Jim brought the plane down through broken cloud and we landed at Sagwon, a gravel airstrip used as a staging post on the Sagavanirktok River. Here we transferred to a Beaver for the short flight to Shrader Lake, ringed on three sides by mountains and glaciers, its ice cover gleaming white in the afternoon sun. On the edge of this magnificent lake we would make our base-camp.

Soon we were on the ice, taxiing on skis with a crunching noise toward the open stretch of water separating us from the shore. It seemed to us to be fraught with risk, but Jim had been at this for a long time and knew that the ice was still several feet thick.

Our fuel supply had been flown in during the winter and dumped on the shore. I knew that helicopters were thirsty, but I was staggered by the size of our fuel dump: it was made up of five-gallon tins, packed in pairs in wooden boxes, and the pile of boxes was the size of a suburban garage.

Ken, the helicopter pilot, had arrived earlier that day with his mechanic and they were sitting on the beach, idly skimming pebbles on the open lead of water. As they saw the Beaver approaching, Ken started the helicopter and prepared to ferry us and the camp equipment off the ice to the shore. A line of one-man tents and the large mess tent soon sprang up along the water's edge, and the survey was ready to begin.

The system for obtaining rights to explore for oil in USA is different from that adopted in most other countries. It is done by auction, or what American oilmen call "land sales." The broad swathe of flat tundra between the Brooks Range and the ice-covered Arctic Ocean – the North Slope as it is often called – was recognized as having some oil potential, especially after oil in small sub-commercial amounts had been discovered in a drilling campaign carried out several decades earlier on behalf of the US Navy. BP had joined with a US oil company called Sinclair – a company familiar to motorists of the day for the dinosaur logo on its "gas stations" – and in1963 the partnership had successfully bid for blocks of land over areas further to our west, around the Colville River.

The Sadlerochit Mountains are a northward salient of the Brooks Range and it was in and around this salient – an area about the size of East Anglia – that we would spend the next two months studying the outcropping rocks: noting their petrology, variations in the thickness of the different strata, their stratigraphic relations, and any indication that they may be potential oil source rocks or reservoir rocks.

There was no chance whatsoever of finding oil in these mountains, but then that was not the purpose of the survey. Our job was to examine the rocks in this part of the Brooks Range to enable the data from seismic surveys across the tundra plain to be interpreted, since it was there that oil might in future be found if it was to be found anywhere in northern Alaska. (Those seismic surveys, incidentally, could only be done in the winter, when the tundra's surface bogs and pools become frozen, and ground vehicles can be used).

If those terms of reference sound similar to the purpose behind my surveys in New Zealand and Libya it only underlines the fact that the science of geology fundamentally underpins the oil exploration industry. When the first hole was drilled for oil in the modern era, in 1859, by an American businessman named Edwin L.Drake, he sited his well at Oil Creek, Pennsylvania, on the simple basis that a natural seepage of oil occurred there. He was lucky and discovered oil which he could pump to the surface in quantities that he could sell. Others have also been lucky with that "wildcatting" approach, generally where oil was already known to be present nearby. But drilling at random for oil would be a recipe for wasting money in astronomic amounts. While the geological data we were about to start gathering would not guarantee that later wells on the North Slope would find oil, the information would certainly dramatically reduce the risk of failure.

The helicopter was the work-horse of the survey – ground vehicles are useless in the summer months and the distances are too great to be covered on foot. Ours was a basic machine, a Bell G-2, with a single Lycoming piston engine, a fuselage which looked as if it was made of Meccano, and a perspex bubble capable of holding the pilot and two passengers. We geologists worked in pairs and on a typical day Ken would drop us in the morning at the start of our planned traverse and pick us up in the evening to fly us back to camp. Finding us in that wilderness of mountains was not always easy but we carried a square-yard of "*day-glo*" cloth which we'd wave when we heard the helicopter in the distance, and this was surprisingly effective, even on the dullest day.

We were working in the heart of the Arctic National Wildlife Range, a pristine reserve which has taken on political prominence in recent years as conflicting lobbies have argued over its future. The wildlife was truly spectacular and provided me with powerful memories. There were occasional sightings of wolf and lynx, and moose frequented valleys where the willow scrub grew thickest. Lemmings were common, and ground squirrels made their burrows wherever the soil was dry enough, their "*siq-siq*" call punctuating the Arctic silence. The bird-life was an ornithologists's dream, as this is the breeding area of many waders, wildfowl, pipits and buntings, and a variety of birds of prey – we would know if we strayed close to a peregrine's nest on some cliff when a sudden loud whooshing noise above would make us duck instinctively as the falcon stooped and missed our scalp by inches.

Early in the survey I was working along a sandy river bed and came across a recent large footprint, perhaps the size of a saucer. My first thought was that it must be a grizzly bear's print and, thinking it may be lurking in the nearby willow scrub, I quickly reached into my rucksack for the Colt 45 pistol we had each drawn from the Company stores. I saw no sign of a bear that day, but a week later Hog-Too the cook, whose job it was to get up first and make the breakfast, called us with the cry "There's a bear in camp!" There had been a light fall of snow in the night and one by one the tent-flaps opened and bleary-eyed men emerged and peered about. Standing at the edge of camp was not one but two bears, a mother and a cub, sniffing the air and probably wondering which tent all the good smells were coming from. As we shouted and jumped about in the snow the two bears turned and, in their own time, slowly ambled off. The first thing I did (when I was sure they were well clear) was go and inspect their footprints. What I saw were enormous footprints, the length of an A4 sheet of paper and almost as broad – those saucer-sized prints I'd seen earlier were obviously just wolf prints.

Night-time visits by grizzlies were not uncommon on our surveys in the following years, the smell of food overcoming their natural fear of man. One morning we found the mess-tent in a shambles. While trying to enter the tent a bear had knocked over the lightweight aluminium table inside on which stood the Coleman cooking stove. The crashing sound must have frightened off the grizzly and it didn't return, but its huge fore-paw marks high on the canvas roof gave away its intentions.

I've not seen the migrating herds of wildebeest on the plains of Africa, but for sheer spectacle the migration of Alaska's caribou in their vast and uncountable numbers must be as heart-stopping. We were by now at a different base-camp, on open tundra in a broad valley – the Ignek perhaps, or the Katakturuk – when the vanguard of the herd appeared in the eastern distance. Their numbers swelled and the entire valley was soon a sea of moving animals, their antlers a forest silhouetted against the sky. As they drew closer we could hear the castanet din of thousands of hooves, and then the herd parted and swept by us like a tidal wave to coalesce again and fade into the distance.

In June, as the snow melts on the tundra and on the lower slopes, the flowering plants burst into life, as colourful and varied as an English meadow. Blue lupins carpet the scree slopes, and Alpines such as saxifrage and mountain aven appear. But this is also the time the scourge of the Arctic takes to the air: mosquitoes. I had experienced mosquitoes on my BSES expedition to Northern Quebec and also as a student in Arctic Norway, but not in these numbers. From the moment we came out of our tent in the morning till the time we went to bed, every one of us had a cloud of mosquitoes around him, a cloud so thick that by clapping your hands you could kill a dozen of them. They were big and hungry for our blood, but mercifully there was a product called *"Off!"* which kept us from being driven mad by them. We carried a can wherever we went, and sprayed head, hands and ankles several times a day. So vital to our survival was this stuff that I can remember one or two occasions of near-panic when Ken dropped us for the day on a mountain, and as he flew away I would reach into my rucksack for the *"Off"* and fail at first to find it. Thankfully, if it was not in one pocket of the rucksack it was probably in another, or else my buddy would lend me a few squirts, and the day's work could begin.

Flying was not possible when the cloud-base was low, as it is vital for the pilot of a G-2 to be able to see the ground at all times; it is the nature of those simple helicopters that if he were to find himself in thick cloud he would be unable to tell whether he was the right way up or upside down – with disastrous consequences. And therefore often the helicopter would be grounded and we would have to spend the day in camp. There were geological tasks to catch up on, or letters home to write, paperbacks to read, and ten-pound trout to catch in Shrader Lake. Or we would simply sit around the mess-tent table and talk. Those grizzlies were a common talking-point – we all shared the same anxiety that one day when rounding a corner we would encounter an angry bear that chose not to shuffle away but to chase us. We had seen on several occasions how fast they could run if alarmed, for example by the sudden appearance of the helicopter. We would discuss at length our different theories for coping with an attack: one might favour standing one's ground and waiting till it was close enough to kill with a burst from his Colt 45; another would advocate lying on the ground, doggo, hoping that the bear would sniff you and then wander away; my ploy would be to run away, discarding rucksack and loose clothing as I went, hoping that he would stop to sniff them for long enough for me to escape. The truth was, of course, that although we did spot a number of bears they were always far enough away that we could give them a wide berth. And while the pistol did provide a crumb of comfort when a bear appeared, they are hopelessly inaccurate in an amateur's hands – and would be even more so if fired over one's shoulder while running away in a state of terror!.

Firearms could be bought at hardware stores in Fairbanks and most male Alaskans had their own small armoury; our helicopter pilots, for example, had a preference for six-shooters which they would wear in Wyatt Earp fashion, slung low with the holster tied by a thong around the right thigh. As well as our Colt 45s we had a rifle which we kept at camp. Target practice became a favourite evening pastime and we developed an easy familiarity with firearms, although I like to think we never lost our respect for them as potentially lethal weapons. We found that a five-gallon can of aviation gasoline with a burning rag beside it would make a good target, and it is not hard to imagine the explosion caused by a direct hit. It was silly behaviour, and probably not in the best spirit of environmental sensitivity, but it was good fun.

Every week or two Jim Magoffin would fly in with fresh supplies. Keeping it fresh was no problem, as beneath the surface cover of grass, roots and rich brown soil, there was solid white ice, and a man with a spade could dig a natural refrigerator in a minute or two. This is the permafrost for which the Arctic is famous, a permanently frozen body of ice that extends for hundreds of feet downwards. While useful as a fridge, the drawback of having ice just beneath the surface was that in camp the ground became squelchy and it was difficult to avoid having perpetually cold feet. It was not till bed-time that we could guarantee being warm, huddled in the finest goose-down sleeping bags atop down-insulated camp beds supplied by the Seattle firm of Eddie Bauer.

Of course, this far north in summer meant that night was a fiction. We would go to bed when the sun was still shining, and even if it slipped below the horizon for an hour or two (as it did by August) the sky was still bright. Most of us quickly became used to this and would sleep the full eight hours without trouble. But Malcolm found it more difficult, the light filtering through the fabric of his tent keeping him awake for half the night. His solution was to make himself a large wood and cardboard structure like a dog-kennel. He placed this contraption on his bed, put the pillow inside, and then at bedtime would slide into his sleeping bag, put his head inside the dog-kennel and enjoy darkness until it was time to get up.

He was an interesting fellow and I spent many days out in the mountains with him. An Ulsterman, this was his first overseas posting and he pined for the fiancée he had left behind in Belfast, a situation made the more poignant by him being a Protestant and his fiancée a Catholic. Perhaps it was those domestic anxieties which were the cause of Malcolm's absent-mindedness and frequent brushes with disaster. On one occasion he and I were climbing a mountain and had to traverse a snow-filled gully. The safest way was to carve steps in the snow, using our geological hammers as ice-axes. I crossed safely, and Malcolm was behind me when I heard him cry out. He had lost his footing and was sliding spread-eagled down the gully. In seconds he had disappeared and I set off as fast as I could down the mountain, looking for him and fearing the worst. Several hundred feet below I found him: no bones were broken but he was badly shaken and his hands were bleeding.

Another situation in the mountains which requires that one pays attention is when disembarking from a helicopter. Often, on being dropped off on some mountain Malcolm would emerge from the helicopter and instead of moving immediately downhill to be clear of the whirring rotor, he was wont to walk unthinkingly uphill. I would bellow at him above the noise of the helicopter: "Malcolm, for god's sake keep your head down!" and he would hear me and duck, narrowly missing being beheaded. No doubt this absent-mindedness was a passing phase, as he went on to enjoy a very successful career with BP.

Len was our cook and he looked after us well. But I must explain how he came by his nick-name, Hog-Too. In the line of tents at base-camp, mine was the one at the end and closest to the mess tent. Len had set up his Colman stove and cooking arrangements in the mess tent and he would wake early every day to set about preparing a hearty American breakfast for us all – normally eggs and bacon with waffles or pancakes and mugs of coffee. Len, poor fellow, had catarrh, and if I wasn't awakened by his clattering preparations close to my tent, then the noise of his early morning throat-clearing would rouse me from my slumbers: "hooooog!" he would go, "hoooooooooog!" I would not have found that so distasteful if it had not been for the fact that after clearing his throat he would spit vigorously, presumably onto the grassy floor of the tent – but I could never be certain it didn't go into the frying pan: "Hooooog-too!"

By mid-August we had completed what we had set out to do and snowfalls were becoming more frequent. From our last base-camp on Alpha Lake an amphibious Widgeon ferried us and our gear to the Sagwon gravel airstrip where we transferred to a DC3 for the flight to Fairbanks.

The first day back in town from an expedition is always a big day. We hadn't seen a bathroom for two months and the first priority on checking into the Travellers Inn was to soak in a bath of hot water, preferably with a cold drink in one hand and music coming through the open bedroom door.

As well as letters from home, there was a letter from Jim Spence at PalosVerdes. Instead of flying back to California with the others, he asked if I would go immediately to help out on the De Long Mountains Survey, as they needed another pair of hands. Although I had been looking forward to returning to the comforts of California, a further fortnight in Alaska was not too great a hardship. The De Long Mountains Survey had started a fortnight later than ours and was examining a coastal area at the western end of the Brooks Range. To get there I had to fly to Kotzebue, and from there a Cessna took me to Jarvis Lake where the party had made its base camp. Later we moved to the coast at Cape Thompson, a disused "distant early warning" or DEW Line station, built to look westwards over the Bering Sea for a possible Soviet attack. Much of our work could be carried out along the sea cliffs, but in other respects it was like the survey I had just completed, relying heavily on the helicopter. The team had a mix of Sinclair and BP geologists among whom were John Martin and Chris Lewis, the latter an avid fan of Robert Service's Yukon poems who was able to recite in full "The Shooting of Dan McGrew" and "The Cremation of Sam McGee" to the delight of the rest of us. Two weeks later the survey was complete and I was back again at Palos Verdes – from one of the Earth's wildest places to one of it's most hedonistic.

I had been expecting to return to England once the summer was over, but without my knowing, Jim Spence had asked London to let me remain in the States for a full two-year tour of duty. The brash exuberance of the place was beguiling and so I agreed without a second thought. I wrote to Anne and suggested she make contact with Personnel Department who would put the travel arrangements in hand; my father took control of finding a tenant for the house in Billericay; and I made the round of Palos Verdes realtors to find a house to rent. A few weeks later Anne and the girls boarded a PanAm jet and headed for California.

Unfortunately, their's was not a happy flight. While walking across the tarmac at Heathrow, Helen became partly drenched by a deluge of jet-fuel escaping from one of the 707's wing tanks. When they had taken their seats Anne changed Helen into a spare set of clothes she had thoughtfully brought along and the flight got under way. But five or six hours into the journey poor Helen complained of her tummy hurting. Thinking that this was just a touch of airsickness Anne tried to distract her, and the remaining hours dragged on. But shortly before landing at Los Angeles Helen's complaints convinced Anne that something else was troubling her. On looking closely it soon became clear that Helen was suffering the effects of the earlier soaking by jet fuel – her lower abdomen was badly burned and she was in considerable pain.

What followed after they landed was a disgrace to the name of PanAm. The family was kept from seeing a doctor for over an hour while the airline did everything it could to distance itself from the accident. I had to take control of the whole matter and organise Helen's attendance at a hospital, and then get her home and into bed. In the days and weeks that followed, the airline was disinterested and off-hand

and it was with great reluctance that they agreed eventually to reimburse the cost of medical attention and Helen's spoiled clothing. When colleagues at the office heard this story they were astonished that I had not immediately engaged a lawyer experienced in compensation claims, as I could have been awarded huge sums in or out of court. I intensely disliked the compensation culture then and I still do, but I dislike as much the shoulder-shrugging disinterest of companies like PanAm.

We took a lease on a house on Indian Peak Road, a modern and typically Californian single-storey house in the district of Rolling Hills. It had a small back yard and like every house in that street the front garden was a blanket of weed-smothering ivy stretching down to the kerb – there was, of course, no sidewalk.

One of our first tasks was to buy a car – there could have been no question of trying to live without one in California where public transport simply did not exist. Car showrooms and used-car lots dotted the landscape but I knew nothing about American cars or what might constitute good value for money. I needed to get the feel of the market and so decided to make a start by visiting some used-car lots in the nearby suburb of Torrance. I went into the first I came to, which advertised itself with the usual strings of bunting and a forest of gaudy signs. As I walked along the first row of cars, I sized them up and considered their age, mileage and appearance. They had no prices on them and before long a brash young salesman appeared with a cheery: "Hi there, how are you today?" We walked along the lines together and I'd stop at one which looked suitable, and ask: "How much is this one?" "That nice Chevy? Two hundred dollars" he'd reply. The next one was, say, a Lincoln and he'd tell me the price was nine hundred dollars. I was finding this quite useful, but I noticed that after about five or six of these stops the salesman was getting restless and impatient. We continued for a little longer, until he could contain his impatience no longer: "What in hell's name are you about, mister?" he expostulated, "first you want a car for $200 and then you want one for $900 … Are you some kindava nut?" And with that he strode away.

I concluded that in California it was more usual for a customer to tell the salesman what his budget was, and the salesman would then show him what he had at that price. At all events, I was put off second-hand cars and instead I responded to TV advertisements and bought a brand-new Chrysler Valiant. And this time the salesman was a model of courtesy.

Helen was now five and we felt she should start her schooling. South Bay Christian School seemed to fit the bill, although Anne was more attracted than I was by its name. Helen pledged allegiance to the flag each morning and was instructed in reading and number, but the school's approach was narrow-minded and austere, and talking in class was punished by a slap with "the paddle". As soon as she was old enough we removed her and placed her in the local primary school where the friendliness and relaxed atmosphere were more to her liking. Sally meanwhile was at a friendly nursery school in Rolling Hills.

There were undoubted attractions to living in California, especially when one lived and worked in such privileged surroundings as the Palos Verdes peninsula. Not for us the daily commute on frenzied freeways to downtown L.A. From the Palos Verdes hills we could look down on the brown smog blanket that filled the Los Angeles basin; and getting to work was a mere few minutes drive. After work we would sometimes have a few holes of golf, which even as a hopeless duffer I'd enjoy, until the evening mist rolled in off the Pacific. The game of squash was new to the States at that time but some expatriates had built a court at the back of a seedy shop in Redondo Beach and I became a keen weekly player. We explored Hollywood and danced to Burt Bacharach tunes at PJ's on Sunset Boulevard. Disneyland was close by, at Anaheim; and that year saw the opening of the new Los Angeles Music Centre, a much-needed architectural focal point in the amorphous L.A. sprawl.

But however great were the urban attractions, I needed communion with the natural world to restore me. At weekends I yearned for the countryside. But to reach it through the vast urban sprawl was more than could easily be done in a single day trip. Once out of the city, the state of California must have some of the finest scenery in the USA. I remember one long weekend before the family arrived, with Mike Savage and Chris Lewis in Chris's open-topped Oldsmobile, camping beside the Colorado River and exploring Joshua Tree National Monument and beyond into Arizona. Another memorable trip was with the family to northern California – to Sequoia and Yosemite National Parks, where we camped in forest clearings beside sparkling mountain streams, and I gave thanks to John Muir, the Scottish-born campaigner who fought successfully in the late 1800s to preserve these special places.

One day at the office I was in conversation with a colleague – a palaeontologist – who, I was intrigued to learn had obtained his PhD from London University as an external student. I was interested that such a thing was possible, and quizzed him on what was involved. It became clear that as a London graduate myself, I too was eligible to submit a research proposal with a view to obtaining a doctorate. Over the days and weeks that followed, I resolved to pursue this idea. My research topic was obvious – the Whangara-Waimata area. That had been my first survey project in New Zealand and I had barely scratched the surface in attempting to unravel its complexity. I wrote to Norman Falcon in Head Office and obtained his approval, then submitted an outline of the research project to London University at Senate House. It was approved, and before long I had enrolled at the library of UCLA and was making plans to return to New Zealand on unpaid leave in early 1966 to carry out more fieldwork. The project took up a great deal of my spare time, working at home until after midnight most nights, and driving to the library in downtown L.A. at weekends. Not for a further year after I returned to England was the work complete and the thesis written. I had it bound and submitted it to the Senate House in the summer of 1967. My examiner was Professor Dan Gill, by coincidence a person I knew well, having escorted him around Tripolitania when he was consulting for BP eight years earlier. I was duly awarded my doctorate.

Spin-off from the further research for my doctorate took the form of two more technical papers. One, a short note to the journal Nature, described the gravity-gliding that I considered caused much of the structural complexity of northeast New Zealand; and the other was a review of mud volcanoes in New Zealand which was published in the Journal of the American Association of Petroleum Geologists, and in which I sought to relate these unusual features to abnormally high underground pressure and, in turn, to the sort of structural complexity which is present there.

At the office Jim Spence assigned me the job of reviewing the geology and the oil prospects of Southern Alaska. At that time, Alaska's only oil production was from fields in and around Cook Inlet – (I seemed to be following in the footsteps of explorer James Cook: first Whitby, then Poverty Bay, now Anchorage and Cook Inlet). In contrast to the North Slope, it was an area of intense oil-industry interest and activity and I was to recommend whether, belatedly, BP should seek an involvement there. Using well-logs (which had passed by law into the public domain) I started to piece together the sub-surface geological picture. Once it became known that BP was considering becoming involved in the Southern Alaska "play", offers from other companies started to arrive on my desk. These were mostly farm-out offers from companies willing to offer a percentage interest in their project in return for BP carrying out a certain programme of work at its sole cost; this may have been a seismic survey or, more often, was an invitation to us to drill a hole to test a prospect they had identified. Since most of the companies had their offices in Anchorage I made many trips there to discuss their offers and to examine their technical data before making my recommendations to BP's management. My first such visit was in October, a few weeks after I had settled Anne and the girls into our house on Indian Peak

Road. The 1964 Good Friday Earthquake had caused massive damage in Anchorage only a few months before. It was indeed as shocking as the newspapers had described. Many suburban houses had been destroyed when the gravels on which they had been built turned to slush with the quake; in some cases they had rotated and I remember seeing upstairs windows that you could now step into from road level. Office buildings in the downtown area of the city had suffered less but some leaned sideways and were fissured from top to bottom, and lines of windows which were formerly straight were now offset.

My Southern Alaska project continued for over a year, but in the summer of 1965 I was asked to break off and take part in another Brooks Range survey. Working again with John Martin and Chris Lewis, we continued surveying from where we had left off in September the previous year, in the western part of the Range. I see from the letters home that my mother kept, we had base-camps on the Utukok River and at Noatak village, a largely Inuit community south of the range. In spite of the frequent fog that rolled inland off the ice-covered sea, the survey was as absorbing and enjoyable as in 1964 – or perhaps more so, as this year I had come north with a small library of books on natural history, to get the most out of the wonderfully rich bird and animal life.

We returned from Alaska nine weeks later to find that the worst racial rioting the USA had ever seen was taking place less than twenty miles from where we lived. The Los Angeles district of Watts, previously known mainly for its eccentrically Gaudiesque Watts Towers, exploded on 11th August into mayhem. Triggered by a minor traffic incident, the pent up frustration over poor living conditions and lack of employment opportunities boiled over, and in the course of the next week 34 people were killed and over a thousand were injured. Predictably, our friends and relatives back in the UK were concerned for our safety, but the reality was that we watched events unfold on our television screens and were as unaffected as they were.

Lyndon B. Johnson was having problems abroad as well. In Vietnam things were going badly for the Americans and troops were being sent there in increasing numbers, with the result that in the States there were campus protests and mounting public hostility to the war. As the holder of a Green Card there was a real risk that I too might be called up for active service. To allow us to work in the States we had entered on Green Cards, which we had obtained from the embassy in London after swearing, among other things, not to overthrow the government or indulge in prostitution. Along with my Bristish colleagues I was ordered to report for the draft, and when some weeks later the letter arrived which would determine my future, I read with relief that in view of my age and my family commitments Uncle Sam did not require my services.

That autumn I was told that in the following year, 1966, I would be running a summer survey on the Alaska Peninsula. The Mesozoic rocks which lie concealed beneath the oilfields of the Cook Inlet occur at the surface on the Alaska Peninsula, and so a study of them there would aid our understanding of these important strata. It is also an area of active volcanoes as well as being home to the world's largest brown bears, and so I was excited at the prospect of getting into the field. Preparing for a major survey was a large job. As well as researching the available geological information there were logistics to arrange – fuel dumps, helicopter contracts, deciding on base-camp locations, how we would be supplied, and by whom.

The work progressed well through the winter and into the spring of 1966, when news of another of Britain's recurring foreign-currency crises began to circulate in the office. It appeared the Wilson Government was warning that the funds needed by BP to continue its Alaskan operations were in jeopardy. Management decided that the Company should refocus its attention on the North Slope since, by that time, we had built up a substantial acreage position there, albeit still of unknown potential. The Alaska Peninsula survey was cancelled and Colin Campbell, the geologist who had been due to arrive from Colombia to accompany me, was redirected elsewhere. I was told that instead I would be taking part in another Brooks Range survey – my third.

As the time to start the survey approached, the news worsened. It appeared that later that year the whole Alaskan venture would be mothballed and the Palos Verdes office would close. In the eyes of top management, and presumably of the UK government, the Company's case had not been strengthened by the failure of our latest well, Colville, which had just finished drilling on the North Slope without finding a drop of oil. I discussed the situation with Anne and we agreed that there was little point in her and the girls remaining in California on their own through the summer, and so they returned to England in May. I terminated the lease and moved out of our house, back into the Eden Roc on Redondo Beach.

As well as my colleagues from the 1965 survey, John Martin and Chris Lewis, we had a fourth geologist join us on this survey, Stuart Buchan. He became a good friend and one I would find myself working with in later years in Southeast Asia as well as in the UK. But getting to Alaska to start the 1966 survey presented an unusual challenge. A strike by airline staff of the domestic carriers that summer prevented us from flying to Alaska in the normal way, and so Stuart and I decided to hire a car in San Francisco and drive to Seattle, from where we were assured we would be able to pick up a notionally 'international' flight to Fairbanks.

We were still somewhere in the fir forests and mountains of Northern California as darkness fell at the end of our first day. Spreading our sleeping bags beside the car, I remember the brilliance of the stars that night as Stuart and I tested each others' knowledge of the constellations and watched a moving pinprick of light which we agreed must have been an early Soviet or US satellite. At the end of our second day's drive we found ourselves in southern Oregon, gazing down into the sleeping volcano which is Crater Lake. We expected to repeat our rough and ready sleeping arrangements here too, by unrolling our sleeping bags beneath the stars, but black bears were prowling around the carpark in search of food, and discretion, we agreed, was the better part of valour as we checked into a nearby motel.

That last survey covered the western part of the central Brooks Range, and we worked from base-camps on the shores of Feniak, Tukuto, Kurupa and Liberator Lakes. Like the surveys I had been on before, the things I remember were the camaraderie, the magnificent wildlife of the Brooks Range and the North Slope, and the pristine mountain scenery. But particularly engraved on my memory are the numerous helicopter incidents we survived.

Flying in the mountains brings particular problems for a pilot: the air turbulence, the frequent low cloud which can block the way ahead, and the landings and take-offs at high altitude where the lift of the rotors is reduced. Over the course of the three summers in Alaska we had some excellent pilots. Often they had earned their stripes as helicopter pilots in Vietnam. Others had been fixed-wing pilots and had converted (with more or less success) to helicopters for the higher pay they could command. But one or two pilots assigned to work with us turned out to be downright dangerous.

At the start of a typical day's traverse we would ask to be dropped by helicopter at the top of a particular mountain. From there we could take in the panorama and sketch the main elements of the geology – and it meant the rest of the day's journey would be mainly downhill! To make a gentle and therefore a safe helicopter landing requires that the pilot approaches the landing site *into* the wind – the higher the altitude and the thinner the air, the more important this is. Whereas airports have wind-socks to show the wind direction, on the rocky mountain tops of the Brooks Range there was not even a blade of vegetation to give a clue, and nothing distinguishes a good pilot from an indifferent one more than his ability to overcome this handicap.

On one occasion that I shall never forget we circled high over our chosen mountain while Gene (not his real name) tried to assess the wind direction. A quick sideways glance showed that he was tense and anxious; his lips were tight-drawn and he gripped the controls with clenched fists. We continued to circle as he strained to work out the best direction for his approach. Finally he made up his mind. We lined up and began our descent. But we were coming down much too fast. He had misjudged it and was approaching down-wind. We were dropping like a stone, the mountain top racing up towards us. In a split second Gene had to decide what to do – he could land on the summit and hope that we and the helicopter would survive the hard landing; or he could try and abort the landing.

He decided to abort. His left hand twisted the throttle and the engine roared, while with his right hand he pushed the stick forward to head off down the valley. But he had left it too late. We hit the far side of the summit a glancing blow and continued over the edge, our skids tobogganing down the scree slope sending rocks flying into the air as Gene fought to gain air-speed. It seemed we would surely come to grief. Helicopters aren't built for high-speed tobogganing and soon we would hit an outcrop and tear the bottom out of the machine, or a flying rock would hit the tail-rotor and shatter it. After seconds, which seemed like minutes, the crashing sound ceased, we were clear of the mountainside and diving into the valley. Having got it so spectacularly wrong, Gene made his next attempt in the opposite direction and we landed without mishap. To say that we scrambled out with a feeling of relief would be an understatement.

On another occasion we made a downwind approach and landed so hard that we felt something must surely have been damaged by the impact. A mountain-top inspection failed to find any damage and, after making our geological observations, we climbed back in and flew back to camp. Only when the helicopter mechanic carried out a full inspection that evening did he discover that our flight back could easily have ended in disaster, as our hard landing had broken a vital cable in the rotor-head – if we had encountered air turbulence, he told us, the main rotor might easily have sliced off our own tail rotor assembly.

Air turbulence was a common problem. I remember one occasion when Jim Spence was away on leave and his place as Chief Geologist was temporarily being filled by Laurie Gay. As part of his familiarization, Laurie decided to pay us a visit in the Brooks Range. He was duly flown in by Jim Magoffin and over the course of the next few days we enjoyed his company and the good-natured arguments he would stir up, as Laurie was a man who held strong opinions on most things. The day he was due to accompany me in the field was a particularly windy one.

That morning the pilot – (I'll call this one Glen) – flew low over the ground to keep below the worst of the turbulence, and dropped Laurie and me uneventfully on a low rocky ridge to begin our day's traverse – no mountain-top landings in weather like that, thank you. All day the clouds thickened and the wind strengthened. By evening the wind had reached near-gale force and we began contemplating a night stranded in the middle of nowhere. But then the distant drone of the helicopter could be heard – Glen, decent soul that he was, had not been deterred by the storm and was looking for us. We got out our *day-glo* cloths and waved them, doubting that he would see us in the dull grey evening light. But he did see us, and with the meagre vegetation flattened by the gale there could be no doubting the wind direction and he landed without difficulty. We crouched, keeping clear of the whirling rotor blades, and scrambled in, strapping ourselves into the bench seat beside Glen. Laurie was in first and took the middle seat, squeezed next to Glen, and I was left with the seat beside the perspex door.

With a wind of such strength, we were airborne in seconds, flying skywards like a just-released kite. Glen struggled to get closer to the ground and soon we were making headway into the teeth of the near-gale and only a few tens of feet above the tundra of the valley floor. Base-camp lay a few miles away and to get there involved climbing the valley toward its head, crossing a low pass, and then following the next valley down until it widened onto more open ground. We settled back in our seats and looked forward to touching down in ten minutes' time. Progress was slow but uneventful until we reached the pass. The headwind was funnelling through it now at gale force and, suddenly, we were in it – an updraft which whirled the helicopter like a ping-pong ball to an immense height. Sweat broke out on Glen's brow and his jaw was clenched as he fought to control the helicopter. It had taken on a will of its own and now, hundreds of feet above the ground, it was being tossed this way and that. When disaster seems close to hand, even the most stoical of passengers seeks something to cling on to, and I clung with all my might to the thin aluminium door-frame. We continued our mad heavenly dance, the engine's pitch changing and the rotor blades cracking the turbulent air as we lurched around the sky. Glen was as worried as we were as he struggled to keep the craft on an even keel, for it seemed intent on turning upside down. If only we could lose height and get below this zone of boiling air .... Eventually, of course, we did. Glen found a down-draft and we plummeted earthwards as fast as we had previously shot upward. As I released my grip on the aluminium door-frame and dared to look around I found that, in his desperation, and with nothing else to cling to, Laurie had found my leg and was hugging my knee to his chest. With an embarrassed expression, he let it go – but whenever we met in later years I teased him about this touchingly affectionate act.

We had one pilot – I think in fact it was Gene – who habitually failed to judge the wind direction. Hard landings and aborted landings seemed to be his stock in trade. We rarely criticised him, even though he frequently nearly killed us, but he realised his shortcomings and over the days and weeks he stayed with us his morale was sapped. One day he was to take us on a long reconnaissance which would allow us to look for rock outcrops projecting from the tundra. It would be over level and open ground, and so flying should be a piece of cake. We set off on a rare sunny day, with me this time sitting in the middle seat. On we droned, a couple of hundred feet up, noting outcrops every few kilometres and scribbling down their details in our notebooks as we went. It was warm inside the bubble and, frankly, the flying must have been tedious. The rock outcrops continued to pass beneath us, but I was increasingly concerned to see that Gene was showing clear signs of sleepiness. The minutes passed. He would lift his cap to scratch his head or rub the back of his neck; he would hold the control column between his knees as he stretched and yawned. He was obviously struggling hard to stay awake. I looked down again. Another rock outcrop lay ahead and I wanted to see what it was – limestone or sandstone, and could I see the way the beds were dipping? But instead of the outcrop passing by beneath, it was coming up toward us – and coming up fast. I looked left at Gene. His eyes were closed, his head lolled forward. He was asleep. In a few more seconds we'd hit the ground, and probably explode in a ball of flames. I'd watched these pilots at close range now for so long that even if I could not fly a helicopter I certainly knew how to make it point upwards. I reached over and grabbed the column from between his knees, pulled it back and the craft came out of its dive. Gene woke with a start: "What the hell are you doing?" he bawled defensively. "You bloody near killed us, you maniac!" I shouted back, my overflowing anger only gradually giving way to relief that we were still alive. As we regained our composure, I warned him that if we came that close to disaster again he would be put on the next plane out to Fairbanks.

Back at Palos Verdes, the survey over, it was a strange feeling to be a part of the wind-down which would lead to the closure of the office. The expatriate staff were dispersing to other posts around the world, and locally-employed personnel – the secretaries and drawing-office staff – were leaving as they found other jobs. Even the contents of the offices and the field equipment went on sale, and I still have a steel straight-edge from the drawing-office which I bought from BP for ten dollars.

Although BP was being obliged to curtail its Alaskan operations, that was by no means the end of the story – in fact quite the opposite. In 1964 BP (without Sinclair, which by then was becoming disheartened by the failure of our joint wells) had acquired at auction a patchwork of acreage around the flanks of a broad dome-shaped geological structure which seismic surveys had revealed in the Prudhoe Bay area. The acreage over the crest of the Prudhoe Bay dome had been won at the same auction (for a higher price) by the American company, Richfield. The recent Colville well, although it had failed to find oil or gas (like the eight earlier dry holes we had drilled), nevertheless provided a large amount of geological data which, when combined with the results of the seismic data and of our geological surveys in the Brooks Range, suggested that this area, around Prudhoe Bay, would be more prospective. This analysis was largely the work of Jim Spence and, in view of the increasingly competitive situation, it was classified top-secret.

That was the position in the autumn of 1966 when the Palos Verdes office closed. What happened less than two years later would transform BP utterly. In 1968, Richfield (by that time merged with Atlantic to form ARCO) announced that it had drilled a successful well on its acreage at Prudhoe Bay. Dollars suddenly became available to BP, which returned to drill on its acreage around the flank of the Prudhoe Bay dome, confirming this to be the largest oilfield in the entire USA. The campaign, which had started in 1959 when BP opened its office for the Alaska venture, was finally successful. The insight of its senior geologists and the willingness of its top management to back its technical staff had finally paid off.

When my time came to leave I took a flight to New York, since to depart the USA without having seen the 'Big Apple' would have been unthinkable. After more than two years in the country I seemed to have absorbed a certain Americanness, for I remember how unfamiliar everything felt as I waited in the British Airways terminal at JFK, among the Brits in their tweed jackets and with their strangely languid voices. And I had forgotten how dismal London can appear, its drab streets of semi-detached houses and parked cars bathed in the grey drizzle of a dull October day.

Chapter Eight

It may have been the era of miniskirts, the Beatles, free love and Carnaby Street, but life back in Harold Wilson's UK seemed rather grim and I was missing the California sun and the 'can do' feel of the USA. Truth to tell, I was also thinking about a girl I had met there after returning from Alaska. She worked in the office next to ours, and my principal memory of that brief liaison is an outing we made together to the Eighteenth Century Spanish Mission of San Juan Capistrano. I remember we shared a bottle of Mateus Rosé, sitting on a breakwater and watching the sun set over the Pacific Ocean. To complete my feeling of alienation, the credit card which I had come to rely on in America was viewed by shopkeepers in England with the same insularity as if I had tried to make purchases in Soviet roubles or glass beads. My spirits weren't lifted when a few days after settling back into our house in Billericay, news of the Welsh coal-tip disaster at Aberfan filled the nation's television screens. More than a hundred children lost their lives when a coal-tip, saturated by the autumn rain, slid onto the village, engulfing houses and the village school.

The pale and tired-looking commuters who shared my train to Liverpool Street of a morning looked even more fed-up when I saw them on the train home in the evening. I daresay that I soon acquired a similar pallor and looked as dejected as they did, but in fact I enjoyed the projects I was given to work on at Head Office.

Following BP's discovery of gas in the UK sector of the Southern North Sea the previous year, the adjacent onshore part of the basin – Yorkshire, Lincolnshire and north Norflok – had taken on a new attractiveness for exploration. I was given the job of assessing the prospects and coming up with an exploration strategy. For the first time in my career I found myself working on a broad front and not a purely geological one. I had frequent meetings with consultants to review their progress on a reservoir-rock study they'd been commissioned to do. There were meetings to discuss commercial matters with our joint-venture partners, British Gas – at which Dennis Rooke participated, later to become BG's Chairman – and with the Canadian company Home Oil. And I had to immerse myself in the UK Government's regulations for awarding exploration and production rights over the areas in which we were interested.

Another project was to re-examine the oil prospects of the Wessex Basin. BP had carried out intermittent exploration there over the years and had discovered the Kimmeridge Oilfield in 1959 – an oilfield which I had myself worked on in the early 'sixties, as I have described – but now a full re-evaluation was required. Geoff Brunstrom was my supervisor, a geologist I knew well as our paths had crossed before in New Zealand as well as in the UK. He was noted in the Company for his precise manner, verging on the pedantic, but I respected his geological acumen. In trying to make sense of the distribution of natural oil seepages along the Dorset coast, and of the failures and the few successful wells that had been drilled, Geoff concluded that the key was to understand the geological structure not as it is today, but as it was back in Createaceous times. The vertically-dipping chalk strata that any walker on that beautiful coastline will have seen – at Lulworth Cove, for example, or on the Isle of Wight – had been deposited horizontally and were up-ended in relatively recent times by the upheaval which created the Alps. By working out the structure as it was before that upheaval – about thirty million years ago – it might be possible to identify places where oil accumulated.

I set about this, and after some months presented Geoff with my report. What was clear was that the major geological fault belt which runs east to west from the Isle of Wight to Purbeck was a long-standing fracture. Most of BP's early drilling had been on the southern fault-block, but it was the northern fault-block, I concluded, which ought to have the better prospects. I felt this to be an important finding and, with several publications already to my name, I wanted to publish it. By that time Peter Kent had taken over from Norman Falcon as Chief Geologist and I sought his permission. A different character from Falcon, Kent maintained close contact with academia and had a long bibliography to his credit – some would say a bibliography swollen by pot-boilers – and so I expected a sympathetic hearing. I was not disappointed, except for his insistence that those parts of my report which might be of commercial benefit to our competitors should be omitted. It was not an unreasonable demand and, of course, I complied, the paper appearing in the Proceedings of the Geologists' Association a year or so later.

Under the terms of our agreement with British Gas, a copy of the report itself was passed to them as our joint-venture partner. Jumping ahead a few years with my story, BP's role as operator for the onshore venture was taken over by Bristish Gas. They carried out seismic surveys in the area and, in 1973, located an exploration well at Wych Farm to test a prospect they had identified seismically. It was on the northern fault block, the geological feature that I had concluded should be the focus of future attention. Their well was a major discovery. The Wytch Farm Oilfield, as it came to be called, is now the largest onshore oilfield in Europe and the operatorship has reverted again to BP.

British Gas undoubtedly deserves credit for making the discovery, although I have had to suppress my irritation over the years when I've heard their Chief Geologist describe how his company did it, with never so much as a passing reference to BP's earlier geological work.

News of the gas discoveries in the Southern North Sea meant that the British public was becoming increasingly aware of the upstream petroleum business (that is, exploration and production as opposed to refining and marketing, the so-called downstream end of the business). From time to time Head Office would receive reports from farmers or gardeners who were convinced that oil or gas was present beneath their land, and they had evidence to prove it, they thought. I was often the geologist sent to investigate.

They were mostly wild-goose chases. An apparently oily smear on the surface of a boggy pool has made many a landowner believe he was going to make his fortune. But I had to tell them that the iridescent film they thought was oil was just a break-down product of chlorophyll in the vegetation. I enjoyed my rail trips around the country, but left behind a trail of disappointed would-be oil millionaires.

One trip, though, was more interesting. A man on Orkney reported that the rocks there contained oil and, in the tiny harbour on the island of Westray, he had seen gas bubbling to the surface. With a young assistant, Martin White, I flew to Kirkwall. It was March 1967 and winter gales were pummelling the North. The next leg of the journey was by ferry through rough seas, and spume was crashing over the ship's bridge as we made our way among the islands. Finally a small open boat landed us on Papa Westray. Sure enough, the flagstones outcropping along the coast did contain traces of black, sticky oil. We eliminated the possibility of contamination, but to make certain we took a number of rock samples for our Sunbury laboratory to analyse. As for the supposed gas bubbling, we saw none – if our informant had seen gas, then perhaps it was from recently buried organic matter on the sea bed. In a week or so Sunbury reported on their analyses. The oil was indeed crude oil, albeit weathered which accounted for its tarry nature. I put all of these findings in a report and concluded that either the oil had been generated within the rocks themselves – the Old Red Sandstone – or else it had migrated into the Old Red Sandstone from younger rocks which subsequently had been eroded away. In retrospect, what made that trip interesting was that at the time the search for oil in the North Sea was still at an early stage and it was not until two years later that the first of many North Sea oilfields would be discovered, and the source of some of that oil was indeed found to have been rocks within the Old Red Sandstone series.

I look back with mixed feelings on the two years that I worked in London. It undoubtedly gave me a broader view of the workings of BP than I had before, and so was an important step in my progress to becoming an oilman. There were some happy times. I started to appreciate what London could offer, and enjoyed fossicking around the historic sites of the City in my spare time. There were regular musical soirées in the intimate surroundings of the Law Society's library, and I remember listening, enthralled, to the guitarists Julian Bream and John Williams play duets. My love affair with boats resurfaced and for £400 I bought a small, second-hand, sailing cruiser which I called *Golondrina*. I spent my evenings refurbishing her in the garage at Billericay and then put her on a mud-berth at Maldon. On Friday evenings after work I'd drive to Maldon with a friend, or on one occasion with Anne and the girls, and we'd pass the weekends exploring the tidal creeks of the Essex coast as I had done as a student. And in the summers we enjoyed holidays with my brother John and his family on the Cornish coast.

But I felt increasingly that Anne's and my marriage was under strain. I first admitted to myself that something was amiss a couple of years earlier, in Alaska. One evening in the mess tent a colleague named Phil had been speaking emotionally about his wife, and about his enthusiasm to return home to her in the 'Southern Forty-eight.' It was, I thought, a typically American thing to have said, and I countered with the observation that for my part the North Slope held as many attractions as California. "Mike" he said increduously, "what's the matter? Don't you love your wife?" It had been a long time since Anne and I had spoken of love, and I lay awake that night turning over in my mind Phil's question. The seeds of doubt over my affection for Anne had been planted.

It was now two years later and the same realisation still haunted me. I found I dreaded the walk to the front door when I returned from the

City each day. We were growing apart in many respects. Our tastes and our attitudes no longer marched in step. And now that she was back in the bosom of her fervently religious family, the beliefs that she adhered to – beliefs that I found irrational and absurd – impinged more on our married life than they had abroad. I increasingly resented the uncompromising Catholic indoctrination of Helen and Sally.

My parents saw that something was troubling me and they did their best to help. But suggestions that Anne and I should "go away and have a nice holiday together" while they looked after the children only deepened my despair. It was not a holiday I needed, but the freedom, or rather the guts, to announce that I no longer loved the woman I had married.

One day in the early autumn of 1968 Joe Glance entered my office. It was the same Joe with whom I'd worked on the Fezzan survey in Libya some ten years earlier. He was still incurably dour, but his taciturnity had mellowed and we now got along well enough together. Managing the career paths of all of the geologists used to be the Chief Geologist's responsibility when there were only seventy of us spread around the world, but as the numbers swelled the task was delegated out to someone else, and for the time being this was Joe. For the last year I had been taking Spanish classes laid on by the Company, thinking that this might encourage the decision-makers to post me next to South America – a prospect I fancied. But that was not to be. Joe sat on the edge of my desk, puffed on his pipe, and told me that I was being sent to Thailand. My knowledge of Asia was sketchy in the extreme and as soon as Joe had left my room, I reached for the atlas. I was required, apparently, to carry out a geological survey around the Gulf of Thailand and I should plan to arrive in Bangkok that October.

That evening I described the day's events to Anne and, since the posting was of uncertain duration, we agreed that initially at least I should go to Bangkok alone. Although leaving Helen and Sally for several months would be a wrench, this enforced separation from Anne seemed to offer the marital breathing-space I needed. A decision on our longer-term future could be deferred to another time.

Chapter Nine

Some air-travellers like to board their plane and remain on board until it gets them to their final destination. As for me, I like to break my journey if there is a chance to explore an interesting place *en route.* And so I decided that I would fly to Bangkok via Moscow, a city I'd not visited before. *Intourist* had allocated me a room at the Istankina Hotel, and I arrived to find a forbidding barracks-like building where a *babushka* was seated on each floor to note guests' every movement. Determined not to be stranded in the grey Moscow suburbs, I found a bus which seemed to be going in the direction of the city centre. With no help from the other bus passengers (all of whom would turn their backs on me when I asked if anyone spoke English), I finally found myself in Red Square. I alighted without even having discovered how to pay the bus fare.

Red Square was as impressive as I had expected. The multi-coloured onion domes of Saint Basil's Cathedral gleamed against the black October night, the red granite Lenin mausoleum crouched ominously, and a large red star shone atop the Kremlin. The cold war was at its coldest and I sensed hostility – or maybe it was just suspiciousness – on the face of almost everyone I saw. There was one exception. I found a restaurant for an evening meal and as I sat there enjoying beluga caviar, borsch, blinis, and a glass of vodka, I noticed a pretty blonde girl watching me from another table. Instead of scowling at me she smiled. Before long we were sitting together and enjoying the rest of the bottle of vodka. But as the evening drew on and I made a move to leave, she said she would like to accompany me. Tempting though this prospect was I had heard about Soviet *femmes fatales,* and images of hidden cameras, entrapment and blackmail swam before my eyes. I had never been with a prostitute in my life and I wasn't about to start with one who, for all I knew, was a KGB agent. I took a taxi back to the Istankina and was happy to sleep alone that night.

I awoke in the early hours, vomiting and with severe diarrhoea. I was sick with food-poisoning. By dawn I was feeling worse, and by repeating *"doc-tor, doc-tor"* to the *babushka* I got through to her that I needed help. She directed me along endless darkened passages to the hotel's clinic. I entered and found a very large, white-clad, Russian woman waiting unsmilingly. With no more than two or three words of Russian I was forced to mime my symptoms – vomiting was easy enough but miming diarrhoea took all my acting skill. Eventually she signalled her understanding. I was made to drop my trousers and bend over and with a syringe resembling a vet's, she administered a massive shot. Miraculously, I was cured within the hour.

An Air India flight took me from Moscow to Delhi, and from there to Bangkok. As I stepped from the 707 the clammy heat of the tropics enveloped me. I felt as though I had been shrink-wrapped inside my damply sticking shirt. My new boss, Noel Crosthwaite, met me at Don Muang Airport and we travelled together in the back of his company car while he explained the office set-up and the arrangements he'd made for my arrival.

The road south from the airport runs across the flood plain of the Chao Phraya River and the view was of rice fields, still under water from the monsoon rains, and criss-crossed by *khlongs*. (Over the next quarter-century that view would be transformed. I drove the road again in the mid-nineties, before the Asian economic crash, and there were no rice fields to be seen; all was factories, offices, suburbia and – oddly – countless used-excavator lots which struck me as typifying Thailand's pell-mell rush to industrialisation.)

Bangkok (or Krung Thep as the Thai people call it) became the capital of Siam in the Eighteenth Century when the long-time enemy, the Burmese, sacked the city of Ayuddhya and the capital was relocated sixty kilometres further south down the Chao Phraya River. Noel's driver wove through the infamously dense traffic. The roads were choked with cars taking city workers home from their offices; with clapped-out Nissan taxis; with buses belching smoke, their conductors making balletic hand signals while they hung from the open doorways; with down-on-their-axles pick-ups loaded high with sugar cane or pineapples; and with the ubiquitous two-stroke, three-wheeler, taxis, the *sam lor* or *tuk tuk*. We finally reached the city centre and I was dropped at the newly-opened Manohra Hotel on Surawong Road, Noel having explained that

BP would get a discount here on account of it's owner being an uncle of Khun Dtao, the office secretary.

At breakfast the next day I had my first lesson in the Thai language. I'd seen Thai writing on signs and hoardings wherever I looked the previous evening as we drove into Bangkok, but I could neither read nor speak a single word. My pretty young waitress repeated the words as I wrote them on the edge of the Bangkok Post I was reading: *malagor, nam som, kafair* – papaya, orange juice, coffee. I said them to myself as I walked to the office, struggling to mimic the tones she had used and which are essential to the language. The office was only a few minutes walk away, and I savoured every step: tailors' shops on New Road offered to make me a suit in 24 hours; every other shop seemed to be a jeweller's; Thai girls in miniskirts were buying orchids for their offices and as they passed me in the street they smiled unselfconsciously and held my eye; but most of all I remember the smells, the cooking smells wafting from a noodle stall a few yards along from the hotel, and sickly smells from uncollected rubbish, and strange smells that I could not identify.

The Thailand venture was a further step in BP's upstream expansion around the world. Since I had joined the Company in 1957 the hunt was on for new areas to explore. With a growing awareness among maritime states that even if they had no onshore oil or gas the prospects off their coasts might be attractive, Thailand's Department of Mineral Resources (the DMR) had offered to the petroleum industry a number of exploration blocks in the Gulf of Thailand. BP was among the companies advised that their applications had been successful and Noel had been sent to Bangkok earlier that year to liaise with the government and establish an office. Whereas BP's British rival, Shell, had a history of Far Eastern upstream operations going back to the previous century, this by contrast was BP's first upstream venture in the region – and I was to be their first technical person on the scene.

The office that Noel had chosen was a small wooden building in Chartered Bank Lane. In fact it was in the shadow of the Chartered Bank itself and it had been with the help of their manager, Norman Eckersley, that Noel had found the premises and taken out a lease. Over the next few years we drew heavily on the bank for administrative help. Noel showed me to my room and introduced me to the only other member of staff, a charming and refined Thai girl named Dtao who had been educated in England. She explained that all Thais have nick-names and hers meant 'tortoise'. And so now I had a vocabulary of four Thai words.

Noel had been an accountant in Head Office and was approaching retirement. He was quite unlike any other boss I'd experienced and I deduced that he'd been selected for this job because of his old-fashioned, courtly manners, which undoubtedly made him a popular figure with the government officials we had to deal with. Although out of a totally different mould from his counterparts in the newly-arrived American oil companies, he got on well with them too and they treated Noel with a mixture of respect and good-natured amusement. I warmed to him immediately and decided that his lack of any technical knowledge would in no way be to my disadvantage.

In the months that Noel had been in Bangkok he had established a wide circle of friends. They included M.R."Bunny" Chakrabandhu, (the M.R. standing for *Mom Rachawong,* meaning that Bunny was a member of the Thai royal family – an extensive group in view of the number of wives that former monarchs had). He lived with his family beside a rural khlong outside Bangkok and I was invited to accompany Noel there on several occasions. Bunny was a delightful man and a generous host who, during the War, had been a leading figure in the resistance to the Japanese occupation. Appropriately, some years later he was chosen to take part in the film "Bridge on the River Kwai," playing the guide who led Jack Hawkins and his sabotage team to demolish the famous bridge.

Since this was Asia, equipping myself with business cards was a priority, and my position was shown as *'Technical Representative, Far East'* – in English on one side and in Thai on the other. Although by now I expected to remain in Thailand for a couple of years, my first job was to carry out a geological survey of the land areas adjacent to the Gulf of Thailand. This meant in effect the whole of southern Thailand below the latitude of Bangkok. An assistant was due to arrive in Bangkok shortly and in the meantime there were survey preparations to put in hand.

We should need Landrovers, of course, and Noel decided that we would hire one vehicle and purchase the other; after the survey was over, that would become my company car, he suggested. Noel's accountancy background ensured he could be relied on to find the most economical solution to any need, and in the case of my Landrover we imported a bare shell and had it modified to a station-wagon in Bangkok. The conversion was carried out in a small back-street workshop and I was astonished at the quality of their coachwork. It was an early glimpse of the entrepreneurial and manufacturing skills which were beginning to change Thailand from being a largely agricultural economy.

The American company, Gulf Oil, had been a long-time partner of BP's in Kuwait and a closeness had developed between the two companies. Although we each had separate rights over different parts of Thailand's offshore area and therefore were competitors, we were nevertheless friendly competitors and early in my stay I was introduced to their Bangkok staff. My opposite number was Paul Truitt, their exploration manager. He too was a geologist and we agreed to make a couple of short trips out of Bangkok before my survey started. One of these was northeast to the hill country where we stayed the night at a guest house in the Khao Yai National Park. We spent the first day examining the geology – cliffs of Mesozoic sandstone with rare fossilised crocodile bones, and ancient rhyolite lava flows – and returned that evening to the guest house for dinner. A group of US servicemen were billeted there and it turned out that they were manning a radar station built by the US military on a nearby mountain top. Inevitably the conversation turned to the Vietnam War, which was at its height. The US servicemen were enthusiastically in favour of LBJ's escalation of the war, and for an hour the argument raged, Paul seeking to persuade his countrymen that the Americans had no right to be there and were propping up a corrupt regime in Saigon. His arguments fell on deaf ears, of course, but I admired and applauded the stand he took.

On the second day of this short trip we had a game of golf at the Khao Yai golf course. Paul acquitted himself well, whereas I went round with my usual score of a hundred plus. We each had a caddy, a pretty young Thai woman, and the most memorable part of the day was when we arrived back at the clubhouse. Having spent a large part of my round looking for balls I'd hit into the rough, my trousers were thick with burs and grass seeds. Before I could even put down my clubs, my caddy was kneeling at my feet with the metal cap of a Coke bottle in her hand, scraping the bits and pieces off my trouser legs. In a few moments she had me presentable again. There are compensations, I thought, for being a golfing no-hoper. Only later did I discover that this simple act of solicitude was characteristic of Thai women's treatment of their menfolk.

Andrew Wainwright arrived a fortnight later. He had just joined the Company and this was his first posting overseas. I could not have wished for a more helpful and able assistant for the coming survey. I had decided that in the six weeks remaining before Christmas we would tackle the southeast corner of the country, between Bangkok and the Cambodian border. The party comprised Andrew and myself, Songserm and Nukun our drivers-cum-field assistants, and Prasai, a young geologist on secondment from the Department of Mineral Resources (the DMR), assigned to us in the name of technology-transfer.

The drive south was across wide rice-growing flood-plains cut by *khlongs,* but as we left the central lowland belt, forested hills ran back from the coast, with fields of maize and tapioca wherever there was level ground. Past the small seaside resort of Pattaya, and through Rayong we drove, and as we progressed southeast the first rubber plantations appeared, and pepper fields resembling vineyards. We stopped from time to time at small towns to drink Coke or buy fresh fruit at bustling open-air markets. And what fruit there was to be had! – hairy rambutan, purple mangosteen, star-apples, custard apples, long or short and sweet bananas, and a host of fruits I'd never seen before. The impression was of food in such profusion that hunger must be unknown. But the towns themselves were marred by streets of ugly concrete shop-houses behind webs of overhead power cables. Only the *wat* or temples with their tall and elegant shape and their gleaming red and green roof tiles stood out against the urban drabness. Our destination was the town of Chanthaburi, and we arrived after dark in a teeming tropical downpour.

On entering an up-country town for the first time the *rong-raem* sign was the first thing we would look for (and this was probably the first Thai writing I learned to decipher). The usual translation is 'hotel' but the word covers a range of accommodation types including, at the bottom end of the scale, a grubby room with a hard and none-too-clean bed and a colony of cockroaches. We established ourselves in a comfortable *rong-raem* in the centre of town.

After the first day working in the field we returned, tired and dirty, to our hotel. We showered and met for dinner before retiring to our rooms, and as I was settling down in my bedroom to work, to plot on the map the day's geological findings, there was a knock at the door. Before I had a chance to investigate, I was astonished when two young Thai women entered and seated themselves on the bed. They spoke little or no English, but it was not difficult to divine their intentions. They were charming and sweet, even though from a business point of view I was a disappointment to them. They seemed to enjoy the contact with a *farang* – a European – and returned most evenings for more laughter and banter, until I would shoo them from the room and lock the door behind them.

The survey progressed well and in a letter to my parents on 20th November 1968 I wrote:

"The region where we have been working is some way inland. It's rolling, forested country, crossed by numerous small rivers and dotted with villages (or moobaan, as the Thais say) of palm-thatched, bamboo or wooden-sided houses on stilts – presumably to keep out the creepy-crawlies and the occasional floods. Chickens, piglets and naked children wander throughout, and in the forest clearings are fields of rice, orange, papaya and coconut trees. These moobaan are linked by a network of dirt tracks which it is our lot to try and traverse by Landrover. It is nothing unusual to find a couple of water buffalo wallowing in a mud pool in the track, which gives you some idea of their surface; on a few occasions we have got bogged down

and have been thankful for the winches. If these tracks don't stop us, then a washed out bridge over some khlong often does.

You meet some pretty colourful people in these back-country parts – little children dozing on a buffalo's back; groups of hunters carrying ancient, muzzle-loading guns; old men on bicycles, sometimes with a pig in a basket on the luggage rack; young men on Honda motorbikes; girls in coolie-hats and carrying billhooks to the rice or banana fields; mahouts on elephants, straining on chains as they haul logs out of the forest; and on the better tracks, gaudily decorated Japanese pick-ups groaning under a load of maybe fifteen locals returning from market.

On the lower, more accessible, country there are extensive rubber plantations, and every household has a conspicuous mangle, used for rolling out the white latex into what look like rubber bath-mats which are then hung out to dry. The tarred road through this wooded countryside could almost be an English country road with its grassy banks, but the illusion is quickly shattered when a coconut plantation or a buffalo comes into view.

In the larger moobaan there is generally a rahn-ahaan where we are able to stop for lunch. They are open fronted with wooden tables next to the road, and on one of the tables will be a line of aluminium pans with various curries, fish soups, vegetable concoctions, and of course a vat of steaming rice over a charcoal stove. The scruffy appearance of these places, where mangy dogs may outnumber the diners, belies the excellence of their food. Andrew and I now eat just what the Thais eat and so far we seem none the worse for it. After several glasses of plain tea or local coffee (strained through what looks like the foot of a woman's stocking – although I'm told it isn't) we set off for the afternoon, usually having bought a water melon or some oranges to keep the wolf from the door till we get back to the hotel about seven.

Andrew is proving a good companion. He's pleasantly easy-going although quite earnest when we approach a rock outcrop. His beard causes great amusement among the children of each village we visit, as the average Thai man has just a few whispy whiskers on his chin. Songserm, my driver, is a great character, always trying to help whether it is pouring me a second glass of tea at the rahn-ahaan or knocking oysters off rocks for me to savour when we're working along the coast. He doubles as my field-assistant and is happy to hammer rocks and label sample-bags, and insists I never carry anything. Nukun is our second driver and is a chap who speaks scarcely a word of English. A bit primitive, and hairy for a Thai, to watch him eating, hunched over a bowl of rice, is quite something! He's a good-humoured chap though. They each get 60 baht a day living allowance, to cover food and lodging, but because they work darned hard I (i.e. BP) always pay their bill when we sit down together for a meal."

A problem I'd not come across on geological surveys in desert or temperate zones was the depth of weathering which affects rocks in this tropical climate. A thick layer of laterite – a hard, brown, iron-rich crust – is present beneath the ground surface which can conceal the true nature of the bedrock and make it difficult to determine whether you're looking at, say, an altered granite or a sandstone. Only along the coastline, where marine erosion has prevented laterite from forming, could we rely on finding fresh, unweathered rocks. And so we spent a proportion of our first six weeks on fishing boats, working our way along the coast, dressed in little more than a sarong and spending the nights sleeping on the deck. Songserm demonstrated his skill at cooking squid on a charcoal grill on deck, and making omelettes from the small oysters we collected off the rocky shore. If there was a more enjoyable way to earn a living, I couldn't think of it. In the same letter home, I wrote of one such cruise:

"We left the hotel at dawn and drove the few kilometres to the fishing village of Laem Sing, one of dozens along this coast. After breakfasting on fried rice and coffee at a waterside rahn-ahaan we boarded our boat and set out to sea. With mist still lingering over the rice fields and clinging to the base of the forested mountains, the view behind us was as beautiful as a Chinese watercolour. Our boat is about 40 feet long, powered by a small diesel and made of stout teak planks secured by teak dowels throughout. Over the flat deck is a temporary canvas awning on bamboo poles to give some shelter from the heat of the sun. Nosing along the coast past empty sandy beaches backed by leaning coconut groves, and past rocky headlands on which grow weird twisted trees, this is truly a wonderful way to do a geological survey. With binoculars we scan the shore for rock outcrops to examine, swimming ashore if it is difficult for the boat to approach. And if there are no outcrops we amuse ourselves by watching the flying fish or focus on an osprey perched in the top branches of a dead tree."

We made one extended boat trip around the forested and sparsely-inhabited islands Ko Chang and Ko Kut, off the southeast coast, and continued to find major discrepancies between the existing DMR published geological map and the rocks we saw on the ground. It was satisfying for Andrew and for me to know that the work we were doing (pleasant as it was) would comprehensively advance understanding of the geology beneath and around the Gulf of Thailand.

After passing one night anchored in a bay on the island of Ko Chang, we found on waking that the stone drinking-water jars on deck were running dry. There was a fishing village at the head of the bay and so the diesel engine was swung into life and we edged toward the rickety wooden jetty. Carrying a couple of buckets I went ashore, certain that I would find a well or a cistern somewhere in the village. I had not walked more than twenty metres when a mangy dog ran at me from beneath one of the wooden shacks, and before I could fend it off it had sunk its teeth into my bare leg. My immediate thought was of rabies. We had been warned of the risk of this disease before coming into the field, and this particular dog had certainly behaved as if it were mad. Forgetting my hunt for drinking water, I ran back to the shore and waded into the sea. For the next five minutes I stood there, rubbing seawater into my wounded leg as if my life depended on it – which I thought it may well have done. Medical handbooks recommend catching the dog and taking it without delay for rabies testing, which, if positive, requires the patient to receive a course of injections in the abdomen. Out in the Gulf of Thailand, and several hundred kilometres from Bangkok, this didn't seem very practical. For the next week or two I worried over my possible foolhardiness, anxiously hoping that I would not see the first tell-tale signs of the disease. The days passed and no one commented that I had begun foaming at the mouth, and in time my fears subsided and I forgot about it.

Toward the end of this nautical odyssey I decided we should cross the stretch of open sea east of Ko Kut, toward the Cambodian coast. Through binoculars from a distance it looked as if the coast was backed by a high escarpment and I thought this warranted a closer inspection. The three Thais in our group, as well as our boatman, were surprisingly nervous at the thought of approaching Cambodian territory and repeated the word *andarrai,* which I learned means 'danger.' But we pressed on, the seas rougher than we had experienced, and our small fishing boat pitched and rolled uncomfortably. As we drew close to shore it became clear that the Cambodian coast was indeed a high escarpment, almost certainly the same group of mostly sandstone beds that forms the rim of Thailand's northeast region, the so-called Khorat Plateau.

We didn't put ashore – I should have had a mutiny on my hands if I'd tried – and so we turned north until we were safely back in Thai waters. The weather remained stormy and the seas rough, and the only haven we could find was a fishing village on the mainland where we tied up alongside a small local factory making fish-sauce. It was only on a cottage-industry scale, but the vats of fermenting fish filled the air with a reek so powerful that we were glad to move on at first light the next day.

The Chanthaburi district is Thailand's principal source of gemstones and we visited several sapphire and ruby mines while we worked there. They consisted of simple holes in the ground where a workman at the bottom would shovel pebbly lateritic soil into a basket and pass it up to others who washed out the potentially precious stones in a stream of water. Of course, most of the stones were not of gemstone quality, but nevertheless it was possible to buy at very reasonable prices handsome stones from the cutting workshops in the town. I bought a pair of star sapphires to make into cuff-links, and blue sapphires and rubies which I had made into a ring each for Helen and Sally when I returned to Bangkok.

By mid-November we had finished our survey in the extreme south-east of the country and were working our way back toward Bangkok. There was strong evidence here of the Vietnam war which was being waged on the other side of the Indochina Peninsula. The Sattahip naval base was a joint Thai-US installation and other subsidiary bases along that coast took advantage of the sheltered deep-water natural harbours. At nearby Utapao was a US airbase from where B52 bombers launched bombing raids on Vietnam. (Only later did the world learn that at the same time Richard Nixon and Henry Kissinger had expanded the aggression and were bombing and napalming Cambodia too). Around Sattahip town a rash of bars and massage parlours had sprung up to cater for the US military personnel.

The main road along that stretch of coast was busy with US military traffic and it was common to see truck-loads of bombs and ammunition shuttling from the naval bases to the airbase and, more poignantly, consignments of military coffins for bringing back the American war dead.

Because coastal rock outcrops tended to be so much better than outcrops inland it was frustrating to find stretches of this coast barred to us for military reasons. We mostly had to accept this but on one occasion, being particularly keen to obtain a key piece of geological information, I gave in to the temptation to enter one of the high-security naval bases. Songserm was, of course, aghast at the idea, but I was insistent. With him at the Landrover wheel, quaking with fear, and me beside him anxiously trying to look as official and as military as I could, we drove up to the barrier. We slowed, and as I looked unblinking at the Thai armed guard he raised the barrier. We drove in. The headland I wished to examine had no buildings nor personnel nearby, and we spent an hour there studying the outcrops and collecting rock samples. We then had to go through the same nerve-racking procedure to get off the base, but again sheer brass neck and arrogance prevailed and we found ourselves back on the public highway, content and considerably relieved.

We completed surveying the southeast corner of the country by mid-December and returned to Bangkok. I moved back into the Manohra Hotel and spent the next fortnight catching up on office work, swimming at the British Club, and enjoying dinner parties hosted by oil company expatriates. Christmas was, of course, not celebrated by the Thai themselves and so passed almost without notice.

With the New Year holiday over it was time to return to the field. Over the next three months I planned to survey the entire Thai peninsula down to the Malaysian border. It was a tall order and we would have little time off, but I was encouraged by our progress over the six weeks before Christmas.

The day before we were due to set off, we assembled at the office to load the two Landrovers. There was not a great amount of stuff: bundles of sample bags, hammers, picks, rolls of maps, water and fuel cans, blankets, and so on, packed into several metre-long wooden chests I'd had made. Songserm and Nukun were there and readily joined in, and our DMR geologist Prasai had been replaced by another, named Prakan. We had one remaining chest to load and as I bent to lift one end I asked Prakan to take the other:

"Hey, Prakan, would you give me a hand, please."

"Sorry, cannot," he replied frostily, standing and looking at me without expression. That's bad news, I thought, perhaps the fellow isn't feeling well.

"Why can't you pick up your end?" I asked, "are you okay?"

"We don't lift boxes. We have labourers to lift them." he explained. I could hardly believe what I was hearing.

"Come on" I said encouragingly, "Andrew, Songserm and Nukun are all busy, so please, come on, give me a hand."

Still he refused. I was now becoming irritated.

"Now listen, Prakan," I spat out through clenched teeth, "I'm going to ask you one more time. I'm the leader of this expedition and I'm willing to lift boxes. If you're not willing to, I don't want you on this survey. Pick up that end of the box, or I shall inform Khun Pisoot!"

Khun Pisoot was a likeable old Thai gentleman at the DMR who was responsible for liaison with the oil companies. Prakan saw that a loss of face would result if I were to carry out my threat, and it may even affect his career prospects. And so with bad grace he lifted his end of the box and we carried it out to the Landrover.

It was not the best possible way to begin a relationship, one in which we would be living cheek by jowl and needing to co-operate to get the job done. But it did give me an insight into the Thai mentality, or more correctly, the Thai male mentality. Frequently over the months that we worked in the field with Thai geologists, we saw instances of the importance they attached to status. They were professional men, they figured, and the proper place for professionals was behind a desk in Bangkok. The reason that Thailand had such lamentably inaccurate geological map coverage began to dawn on me – they were simply unwilling to get their hands dirty by going into the field.

It took four days to drive the twelve hundred kilometres to Songkhla, the main town in Southern Thailand. The road network was still relatively undeveloped and although the main road south along the Andaman Sea coast was mostly tarred it was tortuous where it ran through the mountains; (much of the eastern coast of the peninsula was still accessible only via rough tracks). In a pleasant seaside hotel, the Samila, looking east across the Gulf of Thailand we made our base for the first week or two of the survey, driving out each day to search for outcrops in stream beds or on hillsides.

We had managed to obtain a set of 1:250,000 scale topographic military maps and we worked on, and relied on, these. Week by week, as the colours spread across our geological overlays, a picture of the stratigraphic and structural history of the region started to emerge.

Over the succeeding months we covered the entire peninsula. It was a punishing schedule we had set ourselves and we were under constant pressure to finish before the worst of the hot weather arrived, in April. But what a memorable and unrepeatable experience it was. Unrepeatable because never shall we see again the unspoilt coastline and islands of the gulf and the Andaman Sea as they were before the flood tide of mass tourism, or the vast tracts of rainforest which have since given way to agriculture and industry. Today's holidaymakers in their beach resorts at Krabi or on the islands of Phuket or Ko Samui, enjoying the commercial razzmatazz, the restaurants and night-life, would find it difficult to imagine these places as they were in the 'sixties. Krabi was a sleepy small town and its casuarina-fringed beaches – actually several kilometres away from the nearest habitations – were difficult to access without a four-wheel-drive vehicle and were totally devoid of human life. Phuket had a busy commercial life based on its rubber plantations and on tin-mining, the inclined sluices and muddy lagoons of the latter being a conspicuous feature of the landscape. And Ko Samui lacked a single place to stay and was reached by a wooden ferry-boat on which a First Class ticket for the overnight voyage from Surat Thani cost 22 baht, entitling you to a mattress on the floor of a tiny dormitory cabin.

I still carry in my mind's eye exquisitely beautiful images of those days: a trading junk moored beneath limestone cliffs in the Bay of Phangna; the howl of distant gibbons from jungle-covered ridges as the sun's first rays burned off the night mist; strings of jagged islands like pearls in a crystal blue sea, disturbed only by visits of the bird-nest soup men; darkness falling as we made the up-river approach to Surat Thani after days among the islands, the waterside trees ablaze with the pulsing light of myriad fireflies; gazing down at a flight of hornbills above the forest canopy, as we rested from a long climb on the summit of some mountain peak. But then too there were the heat, the thunderstorms, leeches in my boots, snakes in my Landrover, and some of the grubbiest rooms I ever hope to sleep in where the room-rate was no more than the price of a large bottle of Singha beer.

Surprisingly (considering my poor health later, in Bangkok) we enjoyed perfect health for the whole time we worked up-country. Mosquitoes were an annoyance although, unlike the Arctic mosquitoes, they appeared only after dark. Hardly a day passed without seeing a snake. We carried antivenins in our medical kit, and although we came across cases of snake-bite among those who worked in the rice fields, we remained unharmed. Poisonous scorpions and six-inch long centipedes frequently scuttled away when we turned over a large boulder in the course of our work and we soon learned to avoid putting our hands into places they might lurk. It was more difficult to avoid contact with leeches. The forest was thick with leeches after the rain and while we tried to avoid those we saw, swaying toward us as we approached the twig or leaf on which they perched, it was impossible to avoid those at ground level. Removing your boots after a day in the jungle would reveal blood-soaked socks and several engorged leeches. The very small puncture they left could be slow to heal and, later, when we had returned to Bangkok some of them turned into troublesome tropical ulcers.

Security was a nagging problem in Southern Thailand. Insurgency was taken seriously by the authorities, which mounted patrols to track down the residue of Chen Peng's Malaysian communist party operating in the border area. Communists were the suspected culprits when a railway bridge was blown up while we were in the area, and we heard stories of local people being abducted and even once of foreign aid workers being robbed, but we survived without incident.

Although it was rare to come across other Europeans, we did once meet a pair and it transpired that they were not only British but also geologists. One evening in February 1969 Andrew and I were dining at a roadside eatery near Krabi, enjoying our usual grilled pomfret fish and rice, when a Landrover with a pair of *farang* drew up at the nearby open-fronted hotel. We invited them over for a Singha beer and they explained that they were geologists with the British Geological Survey, Magnus Garson and Andrew Mitchell, and were working on a co-operative project around Phuket. With the painstaking approach for which the BGS is famous, they were close to completing a detailed study of the area's geology and its mineral potential. On several more evenings we met and talked till late, they directing us to richly fossiliferous outcrops they had found, while we were able to explain how the geology changes as one goes east across the peninsula toward the gulf. On one thing we were able to agree: that running Northeast to Southwest across the pronounced bend in the peninsula is a fundamentally important geological fracture, a fault, which had hitherto never been noted. Back in Bangkok some months later, I wrote a technical paper on this fault and submitted the manuscript to Head Office for their approval to publish it. But I had been pipped to the post – Garson and Mitchell's paper on what they named the Khlong Marui Fault had been accepted by the journal *Nature* and appeared in print in early 1970.

By early March we had worked our way north up the peninsula and we spent the last nights of the survey working out of the Station Hotel at Hua Hin. Steam trains rumbled by every few hours between Bangkok and the South, but they brought few new guests and the hotel was struggling to survive. With its airy rooms, heavy brass doorknobs, and mosquito nets hung from high plaster ceilings, the hotel had an old-fashioned charm. It was spotlessly clean and after the grimy places we had become used to, it struck us as luxurious.

On 12th March 1969 the survey was complete and we were back once more in Bangkok. We checked into the Manohra Hotel and that evening sought out the best restaurant in town – the Normandie Grill at the Oriental Hotel. I recall our meal included vichysoisse, chateaubriands, and large wedges of camembert cheese; and we shed no tears that for once rice was not on the menu. The fact that, even back in Bangkok, Andrew and I chose to dine together was testament to how well we had enjoyed each other's company over the preceding four months. Another shared exploit in those first few days back in the city was a visit to a Thai massage parlour. Actually, it was shared only to the extent that we egged each other on and entered the building together. Through a one-way mirror we selected our respective masseuses and met outside an hour or so later, self-consciously admitting that it was a most relaxing and pleasurable experience.

Andrew remained in Bangkok for only a short while before being posted to Turkey. It therefore fell on me to write most of the report on our findings and send it off to London with the set of geological maps we had produced. Copies were deposited with the DMR and I was pleased to see in later years that our maps formed the basis of their published map coverage of southern Thailand, albeit without any acknowledgement whatsoever of the source. So little regional geological work in southern Thailand had previously been done that, undeterred by the disappointment over the Khlong Marui Fault paper, I was prompted to write several publications on other topics on which our survey had thrown new light. I got drawn into a discussion with Garson and his colleagues over the age and origin of a certain group of rocks called the Phuket Group which was published in the pages of the *Geological Magazine;* among its characteristics are its vast thickness and its intervals of pebbly mudstone which prompt debates on whether it was deposited under glacial or deep-sea conditions. And in a paper published in *Nature* I described the Northwest-Southeast faults which we had discovered running along the Moei River valley and through Three Pagodas Pass into Burma. But what really put the cat among the pigeons was a paper I wrote which appeared in *Nature* in 1971 where I proposed that Southeast Asia had formerly been a part of the southern super-continent, Gondwanaland. Geologists working in Malaysia had noted evidence that an unknown landmass had existed west of the peninsula in the geological past, but when I developed the idea, calling on palaeontological and other support, it brought forth howls of protest, in particular from the University of Malaya. I had, admittedly, been wrong about the timing of the "continental drift" which occurred (wrongly suggesting that it was contemporaneous with the collision of India with Eurasia), and it was not until 1980 that I realised that the collision of Southeast Asia with China probably took place much earlier, in Triassic times, publishing the revised paper in the *Journal of the Geological Society.* Nowadays the concept is largely taken for granted.

The theory that the continents have moved relative to each other goes back to the early Twentieth Century, but it was in the nineteen-sixties that the theory of plate tectonics emerged and became widely accepted, combining as it did the earlier notion of continental drift with the new discovery of sea-floor spreading. Through my interest in plate tectonics, which was stimulated by the evidence I saw around me in Southeast Asia, I obtained an insight into – of all things – the Soviet scientific mindset. Based in Bangkok at that time was a United Nations body called the Economic Commission for Asia and the Far East, inevitably shortened to ECAFE, and since geology underpins the economic development of many countries, a number of geologists from around the world were attached to the organization. I got to know some of them quite well, including a Russian geologist named Yuri. We used to talk about what we had each seen of the geology of the region and I soon noticed that he was unwilling to entertain the theory of plate tectonics. In spite of all the evidence, he insisted that the continents had never moved relative to each other – where they are now is where they have always been, he said. Only later did he reveal that he was actually forbidden to believe in the plate tectonic theory. Why? Because the Soviet scientific authorities had not given it their approval.

In early April Anne and the girls came to Bangkok on an exploratory visit. We stayed at the Oriental Hotel for the first fortnight, enjoying its tranquil riverside gardens and its elegant swimming pool. Helen and Sally were soon on the best of terms with the hotel staff and Anne was captivated by the comfort and the unsurpassed standard of Thai service. Noel was due to take home leave later in April and so for the remainder of their stay in Bangkok we moved into Noel's house. It was a large and elegant place off the Sukhumvit Road, with lawns which swept down to a small lily-covered lake, and its mango trees at that time of year were heavy with the most delicious fruit. His excellent Vietnamese cook prepared our evening meals and our other needs were met by the housekeeper and her wash-girl. Anne must have thought she was in Heaven. When in early May it was time for them to return to England we agreed they should join me permanently in Bangkok later that summer, and meanwhile I should find a house for us to move into.

While I scanned the Bangkok Post advertisements for a house to rent, I returned to the Manohra Hotel. Since learning my first three words of Thai over breakfast five months earlier, my grasp of the language had improved and I made friends with the waitresses – girls in their early twenties with a winning combination of physical grace, demureness and a great sense of fun. One weekend I invited a group of them to join me for a day out in the country. We all squeezed into the front of my Landrover and we spent an enjoyable day at a well-known beauty spot northeast of the city, picnicking beside the waterfall at Nakhon Nayok and laughing over each other's linguistic *faux pas*.

There is no doubt that living cheek by jowl up-country with native Thai-speakers had speeded my learning of the language. But I was keen to build on this basic knowledge and I asked Khun Dtao at the office if she could find me a suitable teacher. It was agreed that twice a week after work Khun Buaphan would come to the office and give me an hour or two's tuition. She turned out to be a delightful person, elderly (or so it seemed to me) and clearly a cultivated lady. I realised she was socially from the top drawer during my first lesson when, after a few questions and answers to assess my grasp of the language, she pronounced with shock horror "My goodness, Khun Michael, you speak Thai like a taxi-driver!" Since the little Thai that I could speak had been picked up largely from Songserm, my driver, her comment was more perceptive than she could have imagined.

Throughout my stay in Bangkok, Khun Buaphan continued as my teacher. Thai is a tonal language with similarities to the family of Chinese languages, although much of its vocabulary came, along with Buddhism, from India. Initially, for a European at least, the five tones are a huge barrier to communicating with the local people as to use the incorrect tone is to change completely the sense of what it is you wish to say. For example, to say *glai* with a falling tone means 'near' whereas to say it with a high tone means 'far'. And an amusing sentence which every new student of Thai learns, to drive home the importance of the tones, is *mai mai mai mai?* Provided that the tones used are, respectively, rising, low, falling and high, the words mean 'silk', 'new', 'burn' and the interrogative, and so the sentence becomes 'does new silk burn?' For months I had difficulty with the tones and my attempts at conversation resulted in incomprehension or amusement. I lost count of the times I thought I was ordering a *small* bottle of beer in a restaurant, only to learn that by using the wrong tone I was in fact ordering an *iron* bottle of beer. And so on.

Chapter Ten

Toward the end of the survey I had received a cable from Noel saying that BP in London was taking steps to obtain a foothold in Indonesia, a natural move since this was by far the most oil-productive country in the Far East. Part of the plan was that I should make a visit to Djakarta to talk with the state oil company, Pertamina. And so once the survey was over – in fact a few days after Anne and the girls had arrived in Bangkok – I boarded a Garuda flight to Djakarta. There I was to meet up with David Robertson, the Managing Director of BP's Singapore-based refining and marketing operations, and by a wide margin BP's most senior executive in the Far East. An amusing and urbane bachelor and *bon viveur,* David was the ideal man to head this mission.

Djakarta in 1969 was not the easiest of places in which to do business. Soekarno had been finally toppled from power only three years previously and the city was littered with the rusting steel skeletons of his half-built skyscrapers and grandiose monuments. Telephones worked only occasionally, power failures occurred daily, and taxis were almost non-existent. But the greatest challenge for the visiting businessman was to get a room in the only modern hotel, the Hotel Indonesia. Stories abounded of hotel guests who unwisely dropped off their room key at the front desk in the morning, only to return in the evening to find a stranger asleep in their bed.

By pulling strings among his business and diplomat friends, David had somehow secured rooms at the Hotel Indonesia and it was there that we agreed to rendezvous on Sunday evening. The next day we called at the British Embassy to let them know our plans, and made a goodwill visit to Shell's man in Djakarta, before heading for Pertamina's offices. The principal point of contact with foreign companies was a man called Trisulo, whose title of Technical Director disguised the considerable influence he wielded. He greeted us courteously but was not inclined to let us meet anyone more senior in the organization. This was not enough for David, who insisted to Trisulo that we could not possibly leave Djakarta until we had explained our business in person to Pertamina's chief.

After Suharto, the President of Indonesia, the state oil company's General Ibnu Sutowo was possibly the most powerful man in the country. On that Monday we got as far as his secretary, Mrs Destander. A charming but firm lady, she explained patiently that the President-Director was very busy, but if we wished she would do her best to arrange an appointment for us in two weeks' time. Of course, this was no good for us and David moved into his most persuasive mode, explaining how far we had travelled for this audience, flattering her and her boss, and beguiling her with honeyed words. "All right" she said at last, after disappearing several times into the great man's office, "come back at eight o'clock tomorrow and the General will see you."

BP's exploration history in Indonesia might be said to have begun on Tuesday, 8th April 1969 with that meeting. A very small man, almost lost behind his huge desk, Sutowo sat us down and offered us coffee. "What is this company of yours, BP?" he asked with an expression of puzzlement. Perhaps he really had not heard of it, but whether he had or not, David launched into a polished account of the company and its many successes, finally bringing me into the story and without a trace of irony presenting me as the man who found the Prudhoe Bay oilfield in Alaska before coming out to the Far East. An hour later, with promises that he would make sure we heard about the next bidding round for Production-Sharing Contracts, we were ushered out, thanking Mrs Destander profusely as we went. A few months later a package landed on my desk in Bangkok, describing a number of offshore blocks in Indonesian waters, and inviting BP to submit a bid. With little hard information on which to base a proper technical assessment of the blocks, I compiled a recommendation and sent it off to Peter Kent in London.

In the middle of the year a geophysicist, Mike Ferret, came out from London and joined me in Bangkok. His job was to organise a seismic survey over our blocks in the Gulf of Thailand, and as the results started to come in it was clear that the sedimentary basin – whose existence was essential to the whole Thai venture – was indeed present, at least over large parts of our concessions. It was at this stage in the exploration project that our geological findings were at their most useful, in underpinning the interpretation of the seismic data. Although I was well pleased with the picture of South Thailand's geology which Andrew and I had assembled, one small piece of the geological jigsaw puzzle remained tantalizingly missing. We had managed to visit by boat and collect samples from almost all of the islands in the Gulf, but there was one shown on the Admiralty chart which we had failed to reach. It had the name Ko Losin and lay some 70 km off the coast of Yala Province, close to BP's southernmost offshore concession – which explained the importance I attached to discovering the nature of its geology. The problem was solved in a surprising way. It was at a Bangkok party shortly before I was to go on home leave that I ran into the British Naval Attaché, Commander John Sayer. He was interested in our recent exploits in the South and when he learned of our failure to find Ko Losin he ventured that he may be able to assist. I thought no more about it until a few months later when a large and heavy parcel was delivered from the British High Commission in Singapore. The covering letter was from the Commander of a British submarine which had been on passage through the Gulf of Thailand; it ran:

" ..... *I hope that the enclosed will mean more to you than it does to me. We obtained it during a brief naval exercise. But small wonder you failed to find the island from your fishing boat, as it is only about a metre high and some four or five metres across!*"

It gave me a warm feeling to know that the Royal Navy was able to help the efforts of a British company in this way. The fact that the rock was granite and so proved that this part of our concession lacked any oil potential whatsoever was disappointing but not unexpected, and was at least consistent with the emerging seismic results.

Meanwhile, logistical issues needed to be investigated, issues which were going to arise when our drilling operations got under way. There was Thai Government resistance to the use of helicopters for civil purposes, and radio-communications too were felt could harm the national interest. But gradually the authorities had to accept that if they wished to encourage an oil exploration industry, they could not deny us the tools to do the job. BP's marine experts were sent out to assess possible port facilities in the south of the country from which an offshore drilling rig could be supplied, and in accompanying them I found myself visiting the same small towns I had got to know so well on the geological survey.

As well as a stream of visitors to Bangkok on matters related to our operations, it was a popular place also for BP executives wishing to stop-off when they travelled between Europe and Australasia, and I often found myself entertaining them. Some were happy to have a boat trip up the Chao Phraya River to the Grand Palace, followed by dinner at a good restaurant, but the more disreputable of them wanted to sample the night-life for which the city was getting a reputation. I generally took them to the Café de Paris near Patpong Road where one could dance the night away in the arms of a beautiful Thai hostess. A favourite tune played by the rather creaky band was, I remember, a Barbra Streisand song which ran "People, people who need people are the luckiest people in the world ..." One visitor I remember – a very senior executive from Head Office – disappeared with his hostess toward the end of the evening and did not reappear at the office until just before he was due to catch his plane the next day, an expression of guilty pleasure on his face.

Work continued through the hot and rainy weather of the summer. Bangkok's streets would flood, and on more than one occasion I would return to the Manohra Hotel wading ankle-deep and carrying my shoes in one hand and my briefcase in the other. Part of the job of establishing BP's presence as a serious explorer in the Far East meant that I made frequent trips to oil companies based in Kuala Lumpur and Singapore, and attended various geological conferences. As I got to know David Robertson more, he kindly invited me to stay at his home in Singapore whenever I was in town, and through him I got to know the other BP managers in the region – all 'downstream' people, of course, concerned with refining and marketing. In Bangkok I found a house to rent in a narrow back-lane off Soi Ekamai, and organised a pair of servants who would start work as soon as Anne and the girls returned.

In an age when bookshop shelves sag beneath the weight of guides to every corner of the globe it is hard to imagine that in the late 'sixties there were no practical guidebooks to any part of Southeast Asia. But on one of my visits to Singapore I came across a couple who were working to make good that lack. Staying at the same hotel was a German artist and photographer named Hans Hoefer and his companion, the unusual-named Star Black. They were just back from Bali where they were working on a project which was to result the following year, 1970, in their guidebook to that island – a guide not just to its art and culture but to such practical matters as the places to stay, places to eat and how to travel around. Hoefer settled in Singapore and went on to produce a string of guidebooks in the Insight series. Others followed in his wake, including the Wheelers with their Lonely Planet guides, and now scarcely a town in Asia does not have its humblest hostelry detailed in one guidebook or another.

The Company's policy on home leave was changing at around that time, perhaps reflecting the growing ease and speed of air travel. Instead of leave every two years we were now entitled to annual leave. I intended, of course, to take advantage of the new policy and planned to return to England in August 1969. But before that my colleagues and I were to witness one of history's landmark events. Lacking a television set in our own office we descended on Mollers, the shipping company next-door. They had a set, and there we watched in wonder as Neil Armstrong emerged from Apollo 11's landing vehicle and took Man's first steps on the moon – or so we thought at the time, before the doubters publicised their scepticism.

On the way back to England I stopped in Teheran to see BP colleague, Iain Gillespie, and his wife. They had a house at the top end of town and the thing I remember particularly about my visit is the pleasure of escaping the heat of the day in the Gillespies' pool, from where one could look up through the branches of pine trees at the snow-covered peaks of the Elburz Mountains.

My 1969 pocket diary reminds me that while on leave I took the sleeper train to Glasgow and stayed for a few days with Ian Rolfe, my good friend from school days whom I hadn't seen for many years. We explored the Trossachs and climbed Ben Lomond and I little thought that one day I would return to live nearby.

That year the British Association held its annual meeting at Exeter and Peter Kent suggested I should attend, to wave the flag for BP. All too soon my leave was over and on 24th September, Anne, Helen, Sally and I boarded an SAS flight and headed east again for Bangkok.

We moved into our house on Soi Charoenchai and settled in with Sunee and Chit to look after us. Helen and Sally were enrolled at the Pattana School, an establishment with British staff and run on English lines. Early every morning Chit would stand with them on the footbridge over the lotus-filled creek which bounded our garden and she would teach them words in Thai until the school bus came along. Helen and Sally slipped easily into their new life and adored Chit and Sunee, and when we bought a couple of mongrel dogs from the Weekend Market their happiness was complete. The market, called in Thai *Talaat Sanam Luang,* sprang up every Sunday on the green beside the castellated wall of the Grand Palace. Against a fairy-tale backdrop of gleaming gold spires, stupas and pavilions with curved roofs of sparkling red and green, it captured the essence of Thailand and to wander there was a favourite pastime of mine. In a letter home I wrote:

"Anne and the girls prevailed on me at the weekend to buy a dog. Two dogs actually. After being bitten on the hand while trying to catch a nearly wild puppy I decided it was simpler to buy one, so we

went to the Weekend Market, over the other side of town by the Grand Palace. There we could have bought various fishes, a sea-eagle, a hornbill, hanging parrots, a civet, a baby leopard, mongooses or a small falcon, but we settled in the end on two tiny puppies. One is short-haired, mostly black and very lively. We've named him 'paeng' (which means expensive – because he cost the equivalent of £3); the other is brown and white and fluffy and called 'took' (meaning inexpensive since he cost only twelve shillings). You can imagine how delighted with them the girls are."

The garden of our house on Soi Charoenchai was a source of great pleasure, although my efforts to landscape it had the servants watching with incredulity, particularly over my labours to excavate an ornamental lily pond. They had never before seen a *farang* doing the sort of work which in Asia would normally be done by coolies. My letter home dated 3rd February 1970 :

"Now that it appears we shall be vacating the house, the garden is beginning to look quite good. The banana trees along the back fence are beginning to put out new shoots and so is the frangipani in the tadpole-shaped bed in the lawn. I have embedded granite and limestone boulders in the hillock beside the pond, and the water lily has burst into dark pink flowers. If only we could stop the dogs from wading in the pond. I think they must have some water-buffalo blood in their veins. They are very fit in spite of the tapeworms I keep finding in their faeces. It is no wonder that they get these as they eat everything that comes to hand. Being Thai dogs they eat mostly boiled rice, but soil, elastic bands, grass, paper, string, balloons and cloth are all quite acceptable. I have heard it said that a cure for tapeworms is to place a bowl of gruel in front of the mouth – believed to attract the tapeworm to abandon its host. I'll keep an eye out for a tin of gruel among the Heinz soups when I'm next at the grocer's!

"The garden is alive at present with half-inch long frogs, or more probably toads. Do you remember how our road at Shrewsbury would be covered with them at certain times of year? Also we are getting a lot of warblers on migration, presumably northwards. There seem to be two types: one, the Japanese Great Reed Warbler is about six inches long and sings a repeated "wee, wee, wee", and the other, smaller, one I cannot identify but calls a "see-saw, see-saw" song. The best singer, though, is the magpie robin, a sort of mini-magpie which sits in our half-dead acacia tree.

"The two girls, Sunee and Chit, continue with us and earn every bit of the £30 per month (total) I pay them ..."

My daily routine started early to avoid Bangkok's morning traffic congestion. Leaving home at about 6.30 a.m., I could drive to my office in less than half an hour. Although it lacked the comfort of air-conditioning, my Landrover was a satisfying vehicle to drive and the *took-tooks,* motorbikes and small Japanese cars were inclined to give me a wide berth. From the office it was a couple of minutes' walk to the Oriental Hotel where I would take a table on the terrace beside the river, order coffee, and read the Bangkok Post before starting my day's work.

The British Club was a favourite lunch-time haunt as it was only a walk away from the office, albeit a sweaty walk beneath the noon-day sun. There, after a swim and a spell on the diving boards – where I was better known for enthusiasm than for grace – I'd take lunch beside the pool. That was probably the only place in Bangkok where the lunch menu featured beans on toast, sausages and mash, and fish and chips, firm favourites with the expatriate community. Australians made up a large part of the club membership and I became friends with another lunch-time swimmer by the name of Rod Hood. He looked, sounded, and had the larger-than-life Australian character of a crocodile hunter from the outback, although in fact he was a manager with the Anglo-Thai Company and came from the suburbs of Melbourne. Some years later when I lived in Melbourne, he and his new Thai wife Rudee were to play a prominent part in my life story.

My Gulf Oil opposite number, Paul, also became a close friend and nearly forty years later we still meet in London for a meal with our respective wives, and we have stayed with them at their *hacienda* in the uplands of Argentina. Paul had an Enterprise sailing dinghy which he kept at the Royal Varunha Yacht Club south of the resort town of Pattaya. We sailed together from time to time, and when Paul learned that he was to be posted to Singapore we agreed that I should buy it from him. The long drive from Bangkok was a deterrent to frequent sailing, and in any case the Enterprise is a racing boat and I have always been more a cruising sailor. But I enjoyed those occasional chances to get afloat again.

Around the next headland from the club, in the adjacent bay, King Bhumiphol of Thailand had his weekend sailing retreat. After one Sunday's sailing a lucky coincidence gave me the chance to return home to Bangkok in record time. It was late afternoon and as I drove out of the club grounds I was stopped by a policeman. I sat in my Landrover wondering what the problem was, when around the bend in the road a fast-moving cavalcade of vehicles appeared. Behind several police cars a yellow Rolls-Royce contained the King, and that was followed by more police cars, several Mercedes with the King's entourage, a couple of

Landrovers and, surprisingly, even a small fire-engine. I was waved onto the road, and by accelerating hard I found I could join the tail of the convoy. I have never covered the 140 kilometre journey so fast, and the entire route was lined with policemen who saluted me as I sped by, presumably thinking that in my Landrover I was part of the royal retinue.

By the end of the year, news began to come out of Djakarta that BP was likely to be awarded its first Indonesian block. Lying north of the Mangkalihat Peninsula in a remote part of offshore Borneo, the thinking behind our application had been that a sedimentary basin may be present between the coast and the Muaras and Maratua Reefs. The news of a probable award should have been cause for celebration, but unfortunately it coincided with a time of unusual ill-health for me. On 5th January 1970 I wrote home from the Bangkok Nursing Home:

> *"Only a couple of months after being here with 'flu, I'm back again ....On the plane back from Singapore on the 22nd I felt a bit seedy (and didn't even take advantage of the free wine with dinner), but I put it down to the effects of the BP party I'd attended in Singapore the previous night. Then back in Bangkok I felt progressively worse: on Christmas Day I lazed around and ate almost nothing, and then on Saturday I thought I'd better go and see the quack. He was puzzled but thought that it was most likely dengue fever, picked up from a mosquito somewhere down south. The following day I felt worse – cold shivery spells and then nausea, and then a hot sweat, in about six-hour cycles – so the doctor put me in here. In about three days, with my yellow eyes and distended liver it is now obvious what is wrong".*

I was diagnosed as having hepatitis and was unable to return to work until late-January. I cannot remember whether it was before or after having hepatitis that I was hospitalised again, this time for dengue fever, an ailment which rightly deserves its other name of break-bone fever. And finally, in April, breathing difficulties were diagnosed as being caused by a blockage in the back of my throat, and I returned to the Bangkok Nursing Home where an operation by an excellent Thai surgeon cleared the problem.

I could not have chosen a more pleasant place in which to convalesce than that small cottage hospital on Convent Road. The nurses were gentle and considerate and I enjoyed their ministrations, and we joked that I should get a season ticket, so frequent were my visits. Although located in a busy part of the city there was a large garden behind the hospital where a family of tame gibbons lived in the shade of a huge tamarind tree. As I lay in bed I enjoyed listening to the dawn and dusk calls of these beautiful animals. A Thai legend has been woven around the evocative '*pooah, pooah*' sound of their cry, based on the fact that the word '*pooah*' means 'husband' in the Thai vernacular. The story goes that deep in the forest a young princess was turned into a gibbon when she became separated from her prince, and for the rest of time gibbons have been echoing her call for her lost partner.

Following my recovery I found I was making increasingly frequent visits to Djakarta. Exploration work programmes had to be discussed with Pertamina's staff; commercial specialists would fly from London and we would rendezvous in Indonesia to hammer out our negotiating position on the draft production-sharing agreement; and a library of technical data relating to our new area had to be built from scratch. Many of the early exploration reports on eastern Borneo were housed at the *Institut Teknologi* in Bandung and I made frequent trips to this inland town in the shadow of the smoking volcano, Tankuban Perahu. As photo-copiers had not yet penetrated Indonesia I arranged for these documents to be laboriously re-typed and the accompanying maps to be traced.

It was a hectic and exhausting few months, made more difficult by the inadequacies of Indonesia's infrastructure. On landing at Djakarta's Kemayoran Airport the smell of clove-scented tobacco smoke filled the hot, damp air. It was a heart-sinking sensation because it signalled the start of another period of difficulty, discomfort and frustration. More than thirty years later I can close my eyes and conjure that smell of clove tobacco, and memories of the city and of those arduous visits flood back.

In early March 1970, the ceremony to award BP its production-sharing contract took place. We converged on Djakarta: David Robertson from Singapore, various BP big-wigs from London, and I flew down from Bangkok. Such occasions were still not commonplace in Indonesia and the government was keen to make political capital from our undertaking to spend millions of dollars looking for oil in their country. With Pertamina's top brass and government ministers in attendance, David signed the agreement and the television cameras conveyed the occasion to the nation.

A few weeks later I received news over the oil-industry grapevine that astonished me. It also made me very annoyed. I learned – I believe it was from a BP colleague in Australia – that a young BP geologist was shortly to be assigned to work on a drilling rig off southeast Borneo. My immediate reaction was "That can't be so. We've only just been awarded our Indonesian block, and it is off the northeast not the southeast of Borneo. And besides, it will be several years before we are ready to begin drilling operations there." But the news turned out to be true. Without anyone telling me, BP in London had signed an agreement to join a venture operated by the US company, Union Carbide, over a large block off southeast Borneo. I was very indignant indeed to have been kept in the dark on a matter which so obviously fell within my remit. And I let London know my feelings. To add to my indignation, I recalled that it had been I who had drawn London's attention to this opportunity in the first place. At an oil-industry reception in Bangkok the previous December I had learned in conversation that an existing partner in that venture was looking for another company to 'farm-in', that is to say, to buy its way in. This was exactly the kind of 'scouting' intelligence that I was expected to pass on to BP at head office. I dutifully did so, not imagining for a moment that I would be the last to hear of it if the opportunity was pursued. I never did get to the bottom of this absurd bit of in-house secrecy, but suspect that personal rivalries were at the heart of it.

For me the growing activity in Indonesia meant that it no longer made sense to be based in Thailand where enactment of the Petroleum Law was still awaited and where drilling was therefore still a year or two away. Although I would continue to keep an eye on that venture, the centre of gravity of our operations had shifted south and I began my preparations for a move to Singapore.

The Thailand posting had been a mixture of delights and frustrations. I was completely charmed by the gentle people and by their culture, although it can be a tiresome place in which to live. There is nothing more humanly graceful than a Thai dancer's hand movements, or even the deft delicacy of a kitchen worker as she prepares fresh fruit for the table. And yet Thai bureaucracy and officialdom were as irksome and irrational as any I had experienced. The pollution and the tawdriness of much of the urban landscape could suddenly be replaced by a lush and beautiful garden or the charm of a traditional teak building. And so it was with mixed feelings, but mostly feelings of regret, that I faced the prospect of leaving. From my career's standpoint, the breadth of experience I had gained was hugely beneficial, although from what I could gather my responsibilities were likely to increase further when I got to Singapore.

The delight which Anne had shown when she had visited Bangkok for those few weeks a year previously had evaporated. She was looking forward to home leave and we left on 17th April. On the way to England we broke our journey for a few days of sight-seeing in Lebanon, where we stayed at the Byblos Hotel in Beirut. We had left Bangkok in its hottest season and were poorly prepared for the snowy slopes of The Cedars and the biting winds of the ancient city of Baalbek.

Chapter Eleven

The hardest part of writing these memoirs is recalling and describing the anguish of the next two months. On arrival in England my parents, as generously as ever, allowed us to stay with them at Harold Wood. In an effort to see my life more clearly, again I took the train to Scotland and stayed with the Rolfes near Dumbarton. We went hill-walking and Ian introduced me to the cottage he shared on a remote meadow beside Loch Linnhe. When I returned to Harold Wood, my mind was made up. When I flew to Singapore in June it would be alone, and I would break the news to Anne that I wished to live apart from her.

On 1st June 1970 I left Anne and the girls in a state of great distress with my parents at Harold Wood, and flew to Bangkok. On the flight was a family with two girls of roughly Helen's and Sally's ages and I had to choke back my tears. My girls had done nothing to deserve their father turning his back on them, and yet I knew I could remain no longer married to Anne.

My reason for calling at Bangkok was to examine a yacht that might have been suitable for the geological survey I was planning to run in our northeast Borneo block. She was a 57 foot, yawl-rigged, motor-sailer and undeniably a beautiful vessel, but unsuitable for what I had in mind. I flew on to Singapore and installed myself in the Goodwood Park Hotel. I had agreed with David Robertson that we would establish our exploration office alongside him and his downstream people at BP's offices on Robinson Road, but after a few months it became clear that this was going to be unsuitable for our needs. We needed space for a drawing office and a small laboratory, and as preparations for the geological survey gathered momentum our cramped room on Robinson Road started filling with all of the equipment we would need in the field, including several boxes of paraphernalia I had shipped down from Bangkok. I therefore agreed with David that we should move out from beneath his wing and establish ourselves in the offices attached to BP's bijou oil refinery at Pasir Panjang, on the southwest of the island.

For some time following the decision to establish an exploration presence in Singapore, I had been in discussion with London over the need to build up our technical staff. Initially, Peter Kent expected me to carry out the geological survey of northeast Borneo myself, but I made the case – which he accepted – that if Singapore was to become the centre for conducting not only the exploration operations in the region but also reviewing opportunities to expand our presence, then we should need a substantially larger staff. He agreed to post to Singapore three more geologists to conduct the geological survey: Stuart Buchan was an old friend from my days in Alaska and I was pleased that he was to be the survey leader; Sjoerd Schuyleman would accompany him on the survey and, being a palaeontologist, would remain in Singapore afterwards and set up my planned laboratory; and the third man was to be a young Glasgow geologist who had just joined the company, by the name of Richard Campbell. Meanwhile my research into available ships had brought to light an ex-naval coastal patrol vessel called the *Shalford.* Clyde-built in 1953 and at a charter rate of US $385 per day, she would be ideal for our purposes.

I soon moved out of the Goodwood Park Hotel and into Goldhill Towers, a comfortable and pleasantly airy apartment block on Dunearn Road. From the balcony of my flat on the eleventh floor I could look out over the green, northern, part of the island and watch Malay people coming and going to their nearby *kampong,* the women often stopping to pick a gardenia from the dark green bushes which grew in the gardens below me. After sacking my first *amah* when she announced she was too old and the weather too hot for her to go shopping, I found another, named A-Goon. She had been born in China in 1911 and was of the kind known in Singapore as a 'black-and-white,' an allusion to the traditional baggy black silk trousers and white top she wore over her flattened bosom. She looked after me hand and foot, from the cup of tea with which she woke me at dawn to the excellent meals she had on the table when I returned in the evening. She was a woman of few words and although our communications were on a basic level, I grew fond of her over the time I was in Singapore. On one occasion, however, I did have harsh words with her and it was over a matter of race.

In Singapore, racial harmony was one of the country's most appealing features. At least among the educated classes, the Chinese, Malay, Indian and Europeans mixed on equal terms and I found I soon became oblivious to the race of the people I met through work or at social events. But A-Goon was of the old school and her prejudices were brought home to me one weekend after I had been living there for some months. I had become friends with a family, the Montgomerys. Mike Montgomery was English, and on the weekend in question he had to be away on business. As I often did, I planned to spend the Saturday afternoon at the Swimming Club, and I asked Gina his wife if she and their daughter Jackie would like to come over to my flat for lunch, after which we would go swimming together. They were pleased to accept, and I agreed with A-Goon that she would prepare a Chinese meal for us. At the appointed time there was a knock at the door of my flat and A-Goon padded out from her kitchen and let them in. As she did so, I couldn't help noticing a cool expression cross her face. Gina, a highly intelligent and cultivated woman, happened to be a dark-skinned Indian, and young Jackie was almost as dark as her mother. A-Goon was apparently displeased that I should be entertaining them. I thought little more of her silly behaviour as we talked and enjoyed a pre-luncheon drink, but when it was time for us to come to the table I was surprised and embarrassed to see that whereas my place had been laid with our best white chopsticks, Gina and Jackie had been given cheap bamboo chopsticks that looked as though they were normally used for cooking. Indignantly I asked A-Goon to replace them with white chopsticks immediately. With a grim expression she did as she was told, returning a moment later and slapping down the white chopsticks with conspicuously bad grace in front of my guests. By now I had had enough of her rudeness. I followed her out to the kitchen, closed the door behind me, and let her know that if ever again she insulted my guests she would be instantly dismissed. After that, we lived harmoniously together until it was time for me to leave Singapore.

As an aside I should point out that in Thailand I had found that racial prejudice is as much a part of everyday life as it is elsewhere, notwithstanding the apparent harmony between the races. Those Thais who think of themselves as full-blooded Thai, (many of whom in fact have some recent Chinese in their makeup) look down on the Thais of ethnic Chinese origin and may refer to them disparagingly as *jek*, considering them to lack refinement. And as for the small community of Indians in Thailand, they are looked down on by both of the other races, who have a saying: "If confronted with an Indian and a snake, kill the Indian before you kill the snake."

Within days of my arrival in Singapore in June, Peter Wood also arrived. A drilling engineer by background, he had caught the eye of management in London who considered he would be just the right person to run the new exploration venture in Singapore. Peter was therefore appointed Exploration Co-ordinator and so became my boss. My title remained Chief Geologist Far East, but for at least a year my job had been that of exploration manager and my relationship with Peter was therefore a delicate one. An affable and thoroughly likeable man, Peter had no experience of co-ordinating exploration and although I respected his position and observed the niceties by ensuring he was kept informed of every exploration initiative I took, I felt that he was miscast for this stage of an exploration venture.

On administrative matters my reporting line was up through Peter Wood to David Robertson, but on exploration matters I had two main reporting lines, both of them in London. One line was to Peter Kent, the group Chief Geologist, and the other was to Terry Adams who occupied a position akin to a desk-officer in the Foreign Office. Within a few weeks of my arriving in Singapore, Terry let me know he would like to visit us in order to familiarise himself with the region. A large, bluff Welshman, he had spent the earlier part of his career as a Shell geologist and we got on well together in spite of a tug-of-war which I waged with him over many months concerning where the centre of exploration thinking should be located, Singapore or London. Eventually the decision went my way – or perhaps it is truer to say it went the traditional BP way rather than what I supposed to have been the Shell way.

Terry duly arrived in Singapore and over the next two weeks we travelled together through the region. In southern Thailand I demonstrated the geological picture that Andrew Wainwright and I had unravelled – and introduced him to the hill-billy nightlife of some of the larger provincial towns. We called on government officials in Djakarta and the Technical Institute in Bandung, and we made a trip together to the rig which was drilling an exploration well in the seas off southeast Borneo. This was the well that I had learned about with such annoyance only a month or two earlier, the well being drilled by Union Carbide in which BP had a financial interest and on which one of our young graduates was the resident geologist.

I remember that trip particularly clearly because of a bizarre incident which occurred on the way to the drilling rig. After calling on Peter Quine, BP's newly-installed man in Djakarta responsible for liaison with the government, we flew to the small town of Bandjamasin in southeast Borneo which was the crew-changing base for the drilling operations being carried out offshore in the south Makassar Strait. To reach the rig we had to take the helicopter which made a daily round trip. It was a big chopper, capable of carrying a whole drilling crew, although on this flight we found that we were to be the only passengers. There was room for one of us up beside the pilot and so I insisted that Terry should sit there, since the view would be better and he would enjoy observing the instruments and the controls. I sat in the main body of the helicopter, a cavernous space with seats for about eight. The big sliding door was kept permanently open and I selected a seat beside the opening, facing backwards. After seeing Terry into his aerial perch, I donned my life jacket and from among a tangle of ropes and harnesses I found the right straps to secure myself in my seat – with a yawning opening beside me I wanted to be certain I was well anchored! The engine whined into life, the massive rotor began to turn, and a minute or so later we were rising from the ground and heading east across the mangrove-lined shore. Flying no higher than about two thousand feet gave us a bird's eye view of the shallow sea below, and strapped safely in my seat I was able to lean out through the open doorway and enjoy it to the full. Sharks and giant manta rays swam lazily through the blue waters, here the palest duck-egg blue and there deep indigo as the coral sea-bed changed. There were enormous turtles too, since the atolls and coral islets of the Makassar Strait are their breeding ground.

The drilling rig lay about a hundred kilometres from shore, and although the wildlife below was absorbing, after a while I started to look around the cabin. Twisting in my seat to get a glimpse of the flight deck, I saw that, level with my head and just behind me were Terry's and the pilot's feet. Soon a childish thought came to me. Between Terry's socks and the turn-ups of his trousers, his hairy ankles were exposed. Although I was tightly strapped in my seat and encumbered by my life-jacket and nondescript straps which flapped in the wind from the open doorway, by straining I could just get my hand up to his ankle. In this contortionist's position I selected a hair and pulled it. Down came his hand like a shot, scratching and rubbing what he must have thought was some tropical stinging insect. I chuckled to myself and enjoying the childishness of it I pulled another hair. Again he scratched furiously, and as his hand went up I pulled a third hair. But then, as I was enjoying this silly prank and barely able to stifle my laughter, disaster struck. Without realising it, one of the straps of my seat-belt harness must have flapped out of the open doorway and got caught in the slipstream. In a fraction of a second my body had become gripped as if in a vice and the air was being squeezed from my lungs. The straps had obviously become caught in the whirring tail-rotor, and I was being wound out of the doorway and was about to be chopped into pieces.

In those few seconds, thoughts flashed through my mind. How stupid

I'd been! How paradoxical that I was going to die prematurely because of an infantile practical joke! But more seconds passed, and I was still in my seat and, astonishingly, I was still alive. With a sudden surging sense of relief I saw what had happened. An enormous laugh swept through me – a laugh of nervous relief and amusement at the sheer silliness of the situation. The vice-like grip that I'd thought was going to squeeze the life out of me was caused by nothing more dramatic than my life-jacket's sudden inflation. As I had twisted in my seat to get to Terry's hairy ankle, the toggle had accidentally been pulled. In an instant the built-in gas cylinder had blown up the jacket, and inside my seat harness I was compressed like a tightly trussed chicken. It took the rest of the flight to subdue my laughter and regain my composure. No wonder that on scheduled flights they tell you *"in the event of a landing on water do not inflate your life-jacket until you are clear of the aircraft."*

Readers of that anecdote might form the view that anyone getting up to such pranks could hardly be management material, and who am I to argue with them? Terry, the object of my silly behaviour, went on to become President of BP in Azerbaijan and received an OBE, whereas I ...well, I didn't.

The BP geologist on the rig was a young man new to the company, a Highlander by the name of Donald Milne. He took us to the lab and showed us a log of the strata that had been penetrated so far by the well. It was not encouraging. A few hours later we were back on dry land, on the island of Marsalembo, taking advantage of an enforced stopover to get in some water skiing and scuba diving, both new experiences for me.

A footnote to that story is that the Taku Talu No.1 well which Terry and I visited was abandoned later as a dry hole, as were the subsequent wells drilled on our behalf by Union Carbide. After reviewing the results of the seismic surveys and of the drilling, we concluded that the cover of sedimentary rocks over that part of the Makassar Strait was too thin to have generated oil or gas. In due course, having completed our obligations to the government and to Union Carbide we relinquished our interest in the block.

Back in Singapore I decided that I should take a course of lessons in the Malay language. Although virtually everyone on the island seemed to speak English, I was coming to realise that in my travels through Indonesia and Malaysia a knowledge of the language would be an advantage and would greatly increase the enjoyment I would get from travelling. Bahasa Melayu and Bahasa Indonesia are so closely alike as to be virtually one, and in recent years the two countries have co-operated in removing the spelling differences. BP laid on the course which was attended by a small group of colleagues spurred on by the prospect of a £50 award from the company if we became proficient. Our teacher was a pleasant Malay Singaporean who worked for the broadcasting service. The language has none of the tonal complexity of Thai nor the pronunciation challenges of Arabic, and of course, these days it is written in the Roman alphabet, and so we made good progress. It was also helpful that I had already done a short total-immersion Berlitz course in Indonesian while in London the previous summer. Keen to swell my coffers, I went on from the basic-level to an examination which, I suppose, roughly equated with GCE 'O' Level and for which the Company paid a useful £150 for those who passed.

I was also keen to build on the Thai I had learned, and I was fortunate to locate a charming lady called Mrs Melhuish who agreed to give me lessons. She was elderly and of mixed Thai and European race, and her Welsh husband had died some years previously. Mrs Melhuish loved to talk, and although she spoke perfect English, albeit with the slightly sing-song accent of Singapore, it was in Thai that she chatted to me. Every week she would take the bus along Dunearn Road and appear at my flat in the early evening where she would talk to me solidly for an hour or two in Thai. Listening intently, and straining to follow what she said, I would close the door behind her when she departed and would feel utterly exhausted. But my Thai did improve and I grew fond of the old lady.

After one of my regular Thai lessons I asked Mrs Melhuish if she would care to join me at one of Singapore's many evening food markets for a simple evening meal. She agreed, and I described the incident in a letter home – my parents having become acquainted with the dear old lady by that time:

> *"Mrs Melhuish came round last night for a chat and afterwards I asked if she would care to join me for some noodles down at the Orchard Road Car Park. After walking around the stalls and selecting our meal we settled at a table and I asked her mischievously 'Aren't you embarrassed? You an Asian, being seen with me a European – don't you worry that people will see us and think that you're a social escort or a bar-girl?'*
>
> *I had asked her thinking it would amuse her, but she replied very seriously that she didn't mind what people thought – the important thing is that she knows she is not one of those bad girls. I couldn't help chuckling over this, as she said it with such seriousness. Even A-Goon finds her amusing. As we left home last evening, with Mrs Melhuish talking nineteen to the dozen in Thai, A-Goon caught my eye and gave me a cheeky grin, bringing her hand up to her mouth like a saucy child."*

By early October 1970, after working through a thicket of Indonesian red-tape, all of the preparations for our geological survey of Northeast Kalimantan were complete and the *Shalford* was ready to leave Singapore. The ship was well supplied with food, fuel, fresh water and first-aid equipment and I was not expecting to see the geologists back again much before Christmas. In view of the area's remoteness I knew I would not be receiving frequent progress reports. And so I was concerned when, toward the end of the month, I received a message relayed through a Shell geophysicist who came across the party on the south side of the Mangkalihat Peninsula, that Stuart Buchan had met with an accident and was immobilised. As Stuart was Party Chief, this was serious and so I decided to make my way to the *Shalford* as quickly as I could to stand in for him. The vessel was costing the company nearly $400 per day and I was keen to keep up the survey's productivity.

It seems remarkable now that we had no radio contact with the ship, but that was the case. I therefore had no means of contacting Stuart to learn more about his accident or to alert him to the fact that I planned to visit the party. But before departing, Stuart had told me that he had scheduled *Shalford* to visit Tarakan Island on 30th October, and so it was there that I made for. A bumpy flight on Malaysian Airways got me to the Sabah town of Tawau by evening on the 29th. I took a room for the night at the Royal Hotel and early the next morning boarded a DC3 flight of Bouraq Airways making its regular milk-run across the border to the Indonesian island of Tarakan. *Shalford* was dropping anchor offshore just as the ancient Landrover taxi dropped me at the jetty. An obliging Filipino fisherman took me by skiff the last few hundred yards of my journey. As I climbed aboard, the surprise on the faces of the three geologists gave our reunion something of the flavour of Dr Livingstone's famous meeting with Stanley in Central Africa a hundred years earlier.

I had expected to see Stuart on crutches, but he strode across the deck toward me with hardly so much as a limp. Apparently he had fallen through an open hatch into one of the vessel's holds, badly spraining his ankle but thankfully not breaking any bones as had seemed likely at the time. This meant that for the next couple of weeks we would be mustering not three but four geologists: Stuart, Sjoerd, the fresh-faced young Glaswegian Richard Campbell, and me.

After giving me the good news, Stuart then gave me the bad: a week earlier, while crossing a sand-bar at the mouth of a river the *Shalford* had touched bottom and damaged the port-side propeller. The ship now had a severe judder and the prop needed attention. A conference was held with Dave Ewart, the quietly efficient Australian captain. Although there was a slipway on the nearby island of Bunju it could accommodate only ships upto thirty tons, and *Shalford* was about 130 tons. It was agreed then that we should take the vessel to Bunju in any case to take advantage of the oilfield workshops on the island, but instead of using the slipway we would tie up at a point alongside the jetty where, as the tide fell, she would ground and we could work beneath her.

As we steamed the few miles to Bunju it seemed the vessel would shake herself to pieces. The tide was high and we berthed alongside the jetty without incident. Then it was a matter of waiting for the tide to fall until the water was just a few feet deep when we could go over the side and see the extent of the damage. As the moon came up and the stars appeared, with makeshift lights strung on poles from the stern the problem was all too clear: one blade of the port propeller was badly bent. Half that night we worked upto our chests in water until finally, tired and cold, we were able to haul the bent 350 lbs bronze propeller onto the deck with ropes and pulleys. A truck from Pertamina's workshop took it away the next morning for straightening.

As it turned out, that was the easy part. The field workshop did an excellent job and the propeller was back on the quayside by evening, looking as good as new. Low water was due at about 2.00 a.m. which meant that we could begin work at around midnight. After a hearty meal prepared by Omar, the rotund and jolly Malay cook, we stripped off and soon had the propeller suspended on chains over the stern. To fit it back onto its shaft it had to be manoeuvred precisely and kept exactly in the vertical plane. And that was where the difficulty lay. I described the night's work in a letter home the next day:

> *"It is now Thursday, Guy Fawkes Day. We are still at Bunju, tied up alongside the jetty, but there is not much doubt that this evening we shall be off, steaming south on the high tide. After days of planning and effort, the prop is finally straightened and back on – we laid down the wrenches at five o'clock this morning. The scene last night beneath the stern of the ship was like a medieval vision of Hell: bodies heaving in a mixture of sand and water, straining on chains and ropes in the glare of pressure lamps, with the tide rising till at the end we were working by touch with the water around our ears. A big problem was to keep an open pool of water around the prop and the shaft. We managed this with a hose attached to a petrol-driven pump on deck, but it was a huge battle. As the sand was shifted from one place it would collect in another, and every few minutes we would have to jet our legs free as they became trapped thigh-deep in the quicksand. It was a truly wonderful moment as the threaded shaft began to appear, protruding through the prop. By 5.30 a.m. the job was done and, dog tired, we collapsed into the mess for hot tea and whisky."*

The ship's complement could have been straight out of a tale by Joseph Conrad. Apart from Dave the captain, there was First Engineer Bill (another Australian), a German Australian "Crocodile Dundee"- like character called Fred who sported shoulder-length blond hair and a muscular body the colour of a ripe conker which he cultivated lovingly, two young Indonesian field-assistants, two armed policemen, a Chinese engineer, a Malay deckhand, and of course Omar who was not just a first-rate cook but a man who would happily turn his hand to any job that needed doing. By early evening on the 5th November the ship's normal routine had been re-established and everyone on board looked forward to the engines being re-started and *Shalford* getting under way.

It would be nice to be able to report that with the ship's port propeller straightened, *Shalford's* problems were now behind her. But at sea such mishaps rarely happen singly. Just before high tide Dave started both engines. With the tide at its highest we cast off and inched away from the jetty, everything seeming to go smoothly. Barely had we left the jetty when the engines stalled. Apprehension gripped everyone on board. It was not looking good. The engines were re-started and at low revs *Shalford* crept forward again. In a couple of minutes we were in deep water and Dave gingerly increased speed. As he did so the ship began to shake fiercely from stem to stern and it was clear we had a major problem. The engines were stopped and a couple of us went over the side to see what had happened. The port propeller was unscathed, but the starboard one which presumably had struck the sea-bed and caused the engines to stall had just two blades, the third having broken off.

This was a serious problem which no amount of DIY action was going to solve, and it soon became clear that *Shalford* would have to return to Singapore, 2500 kilometres away, for a new starboard propeller. Normal cruising speed was out of the question, but she could limp along at about seven knots and we decided that *en route* we would investigate the Mangkalihat Peninsula and its neighbouring islands. Again, extracts from my letters home give a flavour of the place.

"For several days we were in the atolls of the Makassar Strait, classic coral reefs with white sand beaches dazzling beneath the hot sun. As we dropped anchor in Maratua Atoll the natives paddled up in their dug-out canoes and swarmed aboard excitedly. Then to Kakabau Island and the tiny islets of Muaras Reef where we met a party of six men from the mainland, out on a turtle-egg gathering trip. In their sailing boat, like an Arab dhow, they showed us a rack sagging under the weight of their haul – no fewer than 6000 eggs, they told us. They sell them for 10 to 15 rupiahs each when they get

back, so they counted their fortnight expedition quite a success. Insisting we need not pay, they gave us fifty of them, soft, white, leathery ping-pong balls which, even when well boiled remain clear and runny – a delicacy in these parts, but not for the squeamish. On the islands we saw countless places where turtles had climbed the beach to lay their eggs, and in the seas are scores of huge adult turtles, as well as porpoises, sea snakes, and teeming small coloured fish. One night, anchored over the reef, we caught two four-feet-long sharks and enough red snapper to fill a couple of fish boxes. The locals in the islands use little wooden goggles for diving, each eyepiece carved and fitted with a _ inch disc of glass cemented in with some kind of resin. We showed one toothless old fellow a modern face mask which we have on board and he put it on. He lay in the shallows with his face submerged, his backside in the air, and then stood up beaming with delight and commenting it was a great improvement on the local product."

"We are now anchored in Sumbang Bay, with hills and jungle rising a thousand feet to the south, east and west, half a dozen palm-thatched houses and a coconut grove fringing the beach. Yesterday, with my field-assistant, Sutomo, we took a villager as guide and tramped inland for several hours through dripping jungle, prickly undergrowth, over fallen trees and through mossy bouldery streams. Finally we reached a small Dayak village of just three houses in a jungle clearing."

"A small, near-naked, old man appeared, with long tresses, a wispy beard and twinkling eyes and beckoned us to ascend the notched log that serves as a stair to his open-fronted bamboo house. His wife and teenage daughter sat there, cross-legged and smiling, bare to the waist, and we passed an enjoyable hour chatting in Bahasa Indonesia with this charming and friendly family. In their small slash-and-burn jungle garden they grow fruit and rice, and what they don't grow they gather or kill with pit-fall traps or with their six-feet long blow-pipes: mostly monkeys and pigs, judging from the skulls and jawbones strung in lines from the eaves. While they may not know how to drive a car or use a telephone their lives are far from simple, as their self-sufficiency must require an encyclopaedic knowledge of birds, animals, insects, plants, natural medications and poisons, the seasons, the weather and the terrain, all passed from one generation to the next. It was a privilege to meet these folk, and well worth the leeches and the struggle of getting there."

Although for me it was the magic of the unspoilt place, its people and its natural history which made the *Shalford* expedition so memorable, it was successful also from the Company's point of view in that Stuart and his party achieved all of the geological objectives that they had set. Using a couple of De Haviland aluminium runabouts with 35 horsepower outboard motors, they made fly-camping sorties up all of the larger rivers of northeast Kalimantan, as well as many of the smaller ones. The geological picture which they were able to piece together undoubtedly helped in the interpretation of the seismic survey which was carried out later.

The decision to move our exploration office to the Pasir Panjang Refinery had been the right one and it was soon running like a well-oiled machine. Once back in Singapore, Sjoerd established his laboratory and a young Singaporean named Joe was taken on to run a cartographic drawing office. Two geophysicists arrived from London to work on the seismic data coming in from the Gulf of Thailand and to plan seismic surveys on our Indonesian contract area. And my team of geologists was joined by Bunny Brown, coincidentally one of the small group with whom I had graduated from UCL thirteen years earlier, although not a close friend. This build-up of staff was evidence that London was coming to accept our argument that Singapore should be the centre of review work on the Far East region, although a final settlement was not reached till mid-1971, at a meeting in London with Peter Kent, Harry Warman and others. Anticipating that decision, I assigned each of the geologists a different area on which I expected him to become expert. I remember that Stuart was assigned on- and offshore Kalimantan, Sjoerd (in addition to running the micropalaeontology laboratory) was to look after on- and offshore Java and Bunny Brown would study Sumatra. I continued to monitor events in Thailand and set about reviewing the South China Sea. Han-Ling Ong, our consultant geologist based at the Institute of Technology in Bandung, continued on a retainer of US$100 per month. Other oil companies had a head-start on us, having had an exploration presence in the Far East for years, and so we had a lot of leeway to make up.

It was about that time, late 1970, that I received a very welcome letter from Head Office. Recognizing the technical achievement represented by the Prudhoe Bay Oilfield in Alaska, the MacRobert Trust had awarded the sum of £20,000 to three of BP's key technical men who had been involved in the lead-up to the discovery: Peter Kent, Harry Warman and Alwyn Thomas. They in turn had generously decided that the prize should be distributed more widely and identified twenty people who had played a part, and I was fortunate to find my name among them. I decided I would spend this £1000 windfall on a trip to Singapore for my parents. Having never been out of Western Europe they were

Fig. 1 York House, Dumfries, in the mid-1930s, where my parents lived in a flat "above the shop" before returning to Essex where I was born.

Fig. 2 Family holidays were taken at Hemsby, in Norfolk, and as a two-year old I was already showing a liking for mud-puddling.

Fig. 3 Final-year geological mapping course, Coniston, Easter 1957.
L. to R: Peter Jones, Len Freddi, MFR, Mervyn Jones, 'Bunny' Brown, Charlie Barnes, Michael Holmes and 'Dusty' Lewis. The picture was taken by Prof. Hollingworth.

Fig. 4 Lunch break with Joe Glance in the El Abiedh oasis, Fezzan survey, Libya, 1959.

Fig. 5 Resting with Ali ben Khalifa on a cliff of red Mesozoic sandstone, Jebel ben Gnema, Fezzan survey, Libya, 1959.

Fig. 6 Hussain Attaila and Ali ben Khalifa bartering for eggs in the market at Murzuk, Fezzan, Libya, 1959.

Fig. 7 At a family party while on home-leave in 1960, I have my arm around my sister, Judy, while on my right stands my mother and on my left, my Uncle Bernard (my father's brother).

Fig. 8 Preparing to land on ice-covered Shrader Lake (left) to set up camp; Mount Michelsen (2699m.) in the distance. Sadlerochit Mountains survey, Alaska, June 1964

Fig. 9 Low cloud grounds the helicopter and keeps us in camp, Sadlerochit Mountains survey, Alaska, July 1964.

Fig. 10 Geoff Larminie observing a herd of migrating caribou, Sadlerochit Mountains survey, Alaska, July 1964.

Fig. 11 Central Brooks Range geological survey party, Alaska, summer 1966.
L. to R. standing: Darryl Coe (helicopter mechanic), MFR, Chris Lewis (geologist), John Martin (geologist), Gene Burlsen (pilot), Fred Walsh (cook); *crouching:* Stuart Buchan (geologist), Hans Spek (field-assistant).

Fig. 12 Snakes abound in Southern Thailand and I found this Banded Krait in November 1968, sunning itself on a jungle track somewhere near the Cambodian border.

Fig. 13 Songserm guides me over a makeshift bridge on a logging track in Southeast Thailand. November 1968.

Fig. 14 About to leave the beautiful granite island of Ko Dtao, having spent the night on the beach under mosquito nets. Early 1969.

Fig.15 Inland from Sumbang Bay I came upon this charming Dayak family, living on what they could grow in their forest clearing and what wild animals the menfolk could catch. Kalimantan survey, late 1970.

Fig. 16 With twinkling eyes, the head of the Dayak family stands in front of his patch of sugar-cane, holding the blowpipe which provides his family with monkey and wild-boar meat.

Fig. 17 At Bundy Hill near Healesville in 1973, wildlife was abundant – including this five-feet long goanna pictured as it climbed a stringybark tree.

Fig. 18 My parents, Don and Stella Ridd, face the camera in the garden of their home at Harold Wood in the mid-1970s.

Fig. 19 Following my recommendation that Britoil should join Arco in its Dubai drilling venture I visited the operation in 1982, and am standing here beside the completed gas well, Margham No.1.

Fig. 20 Shortly after moving into Millfaid we became 'agriculturalists' and built up a small herd of pedigree Highland cattle.

Fig. 21 The long train journey from Thazi to Rangoon, Burma, January 1987.

Fig. 22 Cruising the West Coast of Scotland in *Ashanti* in the summer of 1988.

Fig. 23 John at the helm of *Tethys* in May 1993, on the first leg of our cruise from Itchenor to the Clyde. An exhausted racing pigeon accompanied us in the cockpit as far as Falmouth.

thrilled at the prospect and set about planning for the expedition which we agreed they would make in April the following year.

Although the era of low-cost international scheduled flights had not yet arrived, charter flights were beginning to offer an opportunity to travel inexpensively. As an additional surprise for my parents I decided I would return to England for Christmas. While it was inexpensive, it was anything but comfortable. From Singapore we flew via Bangkok, Bombay and Kuwait before being told the plane would have to divert to Frankfurt. It took another twelve hours to reach London. I remember clearly that flying-visit home.

I recall arriving at my parents' house in Harold Wood and learning from the neighbours that they had just left to spend Christmas with John and his family in Gloucestershire. After making my way to John's house I recall knocking on the door, to be greeted by my mother who visibly staggered backwards across the hallway at the shock and surprise of seeing me, her other son, home from the East. If I had known then what I learned later, that my mother had a heart condition, I would have spared her this potentially dangerous surprise.

After several enjoyable days with John's family and my parents I returned to London where I was welcomed into a warm family Christmas Eve dinner party at the Kensington home of the British Ambassador to Indonesia. I had met Hainworth in Djakarta and again, by chance, in Balikpapan and liked his pipe-smoking and un-stuffy manner. He and his wife had generously entertained me at their Djakarta residence, and I became a friend of their daughter Victoria and had wined and dined her on her occasional visits to Singapore. The Waldorf Hotel on the Strand was a convenient place to stay while in town, and on Christmas Day I awoke to the Christmas-card sight of London lying silent beneath a blanket of snow, putting the seal on my delight.

I spent several days visiting Anne and the girls at Billericay, but, unsurprisingly, relations were not easy. Helen in particular sensed Anne's mood of injury and blame, and reacted with bouts of near-hysteria. Visiting Anne's sister, Jo, widened the hostility toward me, with not only Jo, but other members of the Gallagher family joining my vilification. It was therefore with some relief that I returned to London and the Waldorf Hotel. My last evening was spent with Victoria, who had obtained tickets for a performance of Cyrano, which we followed by dinner at the Petit Montmartre.

That was almost the end of this short interlude, except to add that in the days before it was time to leave I developed severe bronchitis – presumably the result of tobogganing while wearing only light tropical gear. Queuing at Luton Airport to board the charter flight I fell into conversation with a tall slim girl named Doreen who was heading for Hong Kong to take up a position with the broadcasting service. Fate also decreed that we should sit together on the plane, but by then my bronchitis had spread to my larynx and I discovered that I was unable to utter even the faintest sound. We both saw the funny side of this, and spent the entire journey to Kuala Lumpur (where I disembarked) getting to know each other through the exchange of dozens of short written notes.

I cannot leave the Hainworths and their kindness to me in London and Indonesia without relating an incident which occurred when I entertained Victoria on one of her visits to Singapore. We had agreed to have dinner together, and so after picking her up in my car from, I believe, the High Commissioner's residence where she was staying, we drove back to my flat in Goldhill Towers. As usual, I left it in the carpark of the apartment block, not bothering to lock it in view of the bevy of *jagars* who were employed to look after residents' cars and keep them clean. We took the lift to the eleventh floor where A-Goon was on hand to serve us drinks and peanuts as we sat and talked. After an hour or so I suggested we should move on, as I had reserved a table at a smart Chinese restaurant overlooking the harbour. It was, of course, dark by now as I escorted Victoria back to the carpark, opened the passenger door for her to enter, and took my seat behind the wheel. We set off down the long drive and were motoring along Dunearn Road when Victoria let out a sudden fearful scream. Out of the corner of her eye she had caught sight of a figure rising from the back seat of the car. I turned in horror to see within inches of us a man leaning forwards and peering, the whites of his eyes staring at us in alarm out of his black visage. Victoria was naturally terrified, and I myself was somewhat startled. Our unexpected passenger was, of course, one of the Indian *jagars* who, seeing us park the car and apparently retire for the night, did what he normally did and curled up on the back seat, intending to sleep till morning.

It was now 1971, a year that was to see me settle into my bachelorhood in Singapore as well as make a number of business trips out of the country. I had not been back from England more than a week when I was asked by Head Office in London to visit Tokyo to review a farmout opportunity being offered by the company, Japex. The area in question was a block in the Malacca Strait, between North Sumatra and Malaysia, where Japex had carried out some drilling and was now seeking a partner willing to share their past expenditure to earn an interest in the acreage. At the Okura Hotel in Tokyo I rendezvoused with Dave Oliver, one of our London-based geophysicists, and we spent several days at the offices of Japex studying their seismic data and drilling results. It was my first visit to Japan and many of my prejudices were swept away by the hospitality of our hosts who, among other pleasures, introduced us to the delicacy of Japanese food. I had been advised to bring warm clothes with me, and it was as well I did, as Tokyo was gripped by freezing weather. On the Sunday, Dave and I took the train from Asakusa Station in Tokyo for a sight-seeing trip to Nikko, nearly two hours' away. At an elevation of 2800 feet it was even colder than Tokyo. Around Lake Chuzenji the waterfalls were walls of ice, and removing one's shoes to tour Nikko's many beautiful temples and shrines was, I thought, a new kind of foot torture. By the 12th January we had finished our work and Dave and I went our separate ways, he to London and I to Bangkok for lunch with Noel Crosthwaite and brief meetings with friends before returning to Singapore.

Between my frequent trips to different Asian destinations, life in Singapore began to fall into a pattern. At the office we now had a team of geologists and geophysicists able to cover our two principle tasks: planning the exploration work which would lead to drilling on BP's blocks of acreage in Thailand and Indonesia, and reviewing opportunities for BP to expand its portfolio of acreage in the region. Socially, my life revolved around activities with my BP colleagues and my widening circle of Singaporean friends, many of whom I had met through my regular visits to the Singapore Swimming Club. As a young man I had read Lawrence Durrell's Alexandria quartet and now found myself living in a similar milieu of complete unconsciousness to racial differences, a social environment I found most attractive.

BP's expatriate community consisted not just of the exploration team but of a large group of British personnel on the downstream end of our business, concerned with running the refinery, marketing and sales. Among them was a marketing man of about my own age called Dennis Dunstone. He and his wife, Anne, were also members of the Singapore Swimming Club and we would frequently run into each other there and enjoy a meal together. Their young son, Charles, was at the toddler stage and, perhaps missing my own children, I enjoyed playing in the sandpit and splashing in the pool with him. Some thirty years later he founded a company in England called Carphone Warehouse.

As for David Robertson, we had become good but not close friends. After his retirement he settled in London and we continued to meet, sometimes for a drink at his Marylebone flat followed by a meal at one of his favourite restaurants, and from time to time he stayed with us at our home in Scotland. One day he called me to say that he was in hospital. Cancer had been diagnosed and on 7th October 1989 he died, sadly before I was able to visit him. He was a charming and generous man with many friends, but there was something in his nature which kept those friends always at arms' length, a part of him which none could penetrate. Maybe that explains his lifelong bachelorhood, for he never gave the slightest hint that he was homosexual. After his death I, together with another of his friends, Algy Cluff, the mining and oil entrepreneur, presented a small, silver, water jug to his golf club, the Berkshire, inscribed in David's memory.

The Swimming Club had the benefit of being next to the sea, which enabled its Olympic-sized pool to operate on a mixture of fresh water and sea water, without the need for large quantities of eye-stinging chlorine. This was a big attraction which, coupled with its catering facilities, meant the club was a popular recreational venue. I would often swim there after work of an evening and enjoy hurling myself inexpertly from its high diving boards. Frequently my weekend recreation would revolve around the club, but whenever I could I would escape the hurly-burly of Singapore for a weekend with friends, across the causeway in Malaysia. The sleepy town of Malacca, on the west coast, was a favourite, with its old houses and its fort which told of its importance in Portuguese and Dutch colonial history. The empty beaches and jungly islands on the east coast were also a magnet for us, as were the rivers which meandered down to the coast and opened opportunities for exploring by boat the forested hinterland. I recall one weekend trip with Stuart and Jim Hawkins from the office, working up the Endau River in a fishing boat we'd hired at the coast for M$90. We pitched our tent for the night on a wide sand-bar at a bend in the river where we swam, cooked our dinner over an open fire and talked together quietly as darkness settled over the forest and hornbills made their way to their night-time roosts. The next day we pushed on up-river, stopping only when we came upon a village of *orang asli,* the friendly curly-haired aboriginals of West Malaysia.

Back in England my parents were preparing with growing enthusiasm for their trip to Southeast Asia – passports, visas, insurance, tickets, arrangements for Sambo the dog, blank cheques for a trusted neighbour to pay the utility bills – my father was meticulous in such matters. But in a letter he wrote in March, listing progress of their preparations, I also got the first inkling of the descent into hostility which was to mark my later relations with Anne. He wrote:

> *" ...I'm afraid Anne's attitude toward Mother and myself has become so cold that you would think we had both done her some grave hurt. Goodness knows, we have tried in every possible way to maintain a normal relationship but she will have none of it, and her concerns are now so mercenary that spite has taken control.*

> *"I think it is as well that I should explain the position now as we don't want this to be the main topic of conversation whilst we are with you to put any shadow on your and our happiness. It is my opinion that Anne has recently been getting advice from plenty of 'armchair lawyers' as to her present and future position, and she now uses the word divorce quite freely and talks of the financial settlements (quite exaggerated I'm sure) which this would involve. ...It is all a great pity and very saddening. Do not blame the girls for not writing. Whilst I am not prepared to say they are actively discouraged from doing so, I will say with certainty that they are not encouraged. I asked them both today if they would like to write notes for you to be included with this letter, but as they both said they would have to ask Mummy first, and did, the result is obvious."*

There were, of course, no enclosures with that letter.

In April my parents arrived, having broken their journey in Bangkok where a Thai friend took them under her wing and showed them the sights of the Grand Palace and the floating market. They stayed with me for around two weeks and their visit gave them topics of conversation for the rest of their lives. Years later my father would tell of the chilli crab we ate at the food-stalls which in those days sprang up every night in the car-park opposite Robinsons department store, or of the time BP put its launch at our disposal for a languid day of eating, drinking and swimming among the reefs and islands in the Strait of Singapore. My father, normally so conservative in his tastes and attitudes while at home, seemed to delight in these new experiences, from my brown and olive-skinned friends to the wide range of exotic foods we sampled. Perhaps he would have explained this paradox by saying that Oriental things are fine in their proper place – the Orient.

As the east coast of the Malay Peninsula had become a favourite haunt of mine, I decided that an extended trip there over the Easter holiday would be enjoyed by my parents. We left Singapore after work on the Thursday evening and it was dark by the time we were across the causeway. A propitious beginning to our expedition was to catch a full-grown elephant in the car's headlights beside the road as we rounded a bend on the way to our first night's stop, at a government rest-house by the beach at Mersing. The road north was interrupted by ferries at Marchang, Rompin and Endau, where wide rivers which flow off the granite Main Range enter the South China Sea, and so our progress toward our destination, Kemaman, was slow. But it gave us ample chance to linger at Malay kampongs along the way, simple villages of palm-thatched wooden houses where we could buy fruit, or drink fresh coconut milk, and I could practice my *Bahasa Melayu*.

It was late afternoon when we reached the Kemaman Motel where I had booked a triple-bedded room. Situated a stone's throw from the water's edge, with swaying palms and golden sands stretching into the distance, this was the kind of tropical paradise my parents had been looking forward to. We sat for a while in the lengthening shade of a casuarina tree before I decided to strip to my underpants and enjoy a cooling dip in the sea. My mother, watching me cavorting in the waves, was keen to join me, and said to my father: "Don, be a dear would you, and pop up to our room and bring down my swimming costume?" From her childhood, my mother had been an enthusiastic and competent swimmer, although it was probably twenty years since she had been in the water. She had bought a new costume from Marks and Spencer for this holiday and now she waited with growing anticipation for the chance to slip it on and join me in the sea. "Where has Dad got to?" she wondered aloud when my father failed to reappear after some ten minutes. More minutes passed and she could stand the waiting no longer. In an act which astonished me and amused me in equal measure, she stood up, dropped her trousers and shed her blouse, and ran into the sea to join me – a sixty-something year old lady in bra and pants, laughing and frolicking in the sea like a teenager. Now that she has gone, it's a scene that I recall fondly.

At work I was continuing to look for new exploration opportunities where I considered the balance of political stability, accessibility, fiscal terms and geological prospects, might be appropriate for BP to pursue. I advised Head Office that South Vietnam, then still at war, had announced it was offering blocks of acreage to oil companies and I proposed that I should visit Saigon to make contact with the state oil company and gather such data as I could. My proposal was turned down, which disappointed me as I felt that there was an excellent opportunity to enter an oil "play" which was at the time open only to non-American companies.

For some time I'd also been trying to make a case to London that we should get a foothold in southern Sumatra. It had a history of oil production going back to the early part of the century and I thought there were still opportunities to acquire prospective acreage. I had assigned Bunny Brown to carry out a desk study of the island, but now I decided that I needed to make a visit on the ground, to see the rocks at outcrop and assess the operational conditions.

I allowed myself a fortnight for the trip and left by plane for Djakarta in late April. Peter Quine, our representative there, had organized a vehicle for me as well as a jovial, rotund, self-confident driver named Amir and an assistant whose name seemed to be Mothball, a small and wiry fellow with a drooping moustache and a forlorn air. After a couple of days in Djakarta, meeting joint-venture associates and the people at Pertamina, the mini-expedition set off, heading for the small ferry port of Merak at the western tip of Java. Our vehicle was a military jeep left by the Soviets after their failed attempt to establish a lasting presence in the country in the early 'sixties. Like Soviet vehicles I've been in since, it was an uncomfortable but wonderfully tough machine.

That night we crossed the Sunda Strait on a ferry noisy with animals and chickens and a cross-section of Indonesian society. The three of us shared a hot and airless cabin bearing the copious stains of previous travellers, and Mothball passed the night groaning with stomach trouble.

We landed on Sumatra at the small town of Teluk Betung and based ourselves in the scruffy Hotel Kentjana, making day trips on roads so deeply-rutted that in places it was better to drive off them. We stopped wherever there were rock outcrops to be examined and after several days checked out of the hotel to enable us to penetrate further inland. Through Mandah, Terbangi Besar and Menggala to Ogan in the hills southwest of Kota Bumi. That evening in Ogan we found ourselves with nowhere to pass the night, but undaunted, Amir set out on foot to find the *kepala kampong*, the village headman, returning fifteen minutes later to announce I had shelter for the night. It was a simple *attap* kampong house and the family looked after me like their prodigal son. There was something magical about the next day's breakfast: fresh papaya, local coffee and boiled eggs while seated on their veranda, with the sound of cockerels crowing, the sun burning off the overnight dew and the tropical village stirring into life – I can remember it clearly to this day.

Entering the village of Kota Bumi a couple of days later, we saw there was a large gathering of excited men clustered around something on the ground. Curious to know what interested them, I approached and edged my way forward. I heard the muttered word *harimau* and as I reached the front of the crowd I found what so excited them. Slung by its bound feet from a bamboo pole was a full-grown Sumatran tiger. It was dead, of course, and a canvas sack had been pulled over its head, to protect onlookers from the evil eye, I was told. It had been killed in the jungle that morning, but whether for its supposedly medicinal body parts or because it was menacing the village I couldn't discover. I got back in the jeep and we drove on, but I was saddened by the thought that now there was one fewer of this most endangered of the tiger subspecies.

The Barisan Mountains are a volcanic chain with peaks over twelve thousand feet which runs the length of Sumatra, and I was keen to cross it to see the southwestern coast of the island. Through Martapura and Baturadja to Kampong Penjandingan we drove, staying again at the house of the *kepala kampong.* If the lowland roads we had travelled were bad, the road over the mountains was much worse. It was no more than a rough track formed of angular, black, andesite boulders. Past the beautiful crater lake at Ranau and on to Liwa, the last thirty kilometres of the road to the coast were the worst we had come across, resembling the bed of a mountain stream. Every metre of the way had to be won, and large boulders repeatedly threatened to tear the bottom of the jeep from under us. We were on the downhill run when the brakes failed. Brake fluid was dripping from a ruptured pipe and it was obvious it had been severed by a boulder. In a masterly show of make-do-and-mend Amir pushed a nail into the broken end of the pipe, hammered it tightly shut, topped up the fluid reservoir, and we drove on down toward the Pacific Ocean with the brakes working again on the other three wheels.

Finally we reached the coast and the tiny town of Krui. It had about it an air of Arabia and was well worth the journey. Two or three brightly painted fishing boats were pulled up on the beach, and among a cluster of palm-thatched houses we found a charming small *losmen,* a simple, traditional, Indonesian inn where we spent a peaceful night, lulled by the sound of the Pacific surf. The forested mountains we had crossed were well named *Barisan,* as in Indonesian it means 'front' in the sense of being an unyielding line or block (as in *Barisan Nasional,* the Malaysian political party). This corner of Indonesia received few visitors, and those who did reach here presumably arrived by sea rather than across the mountain range. Frequently in my travels I have counted myself fortunate – even privileged, if that is not too clichéd – to find myself among people who had little experience of the outside world, and the feeling was rarely stronger than in Krui. Here they even had their own dialect, and my basic Indonesian fell far short of what was needed to converse.

The same ferry, the *Bukit Barisan*, took us back across the Sunda Strait but the return journey was by day. An hour out we passed close to Anak Krakatoa, the smoking and dust-belching island cone that is growing from the roots of its famous parent, Krakatoa. That volcano exploded one August day in 1883 with a force estimated to have been over 7,000 times greater than the bomb dropped on Hiroshima. Since then the 'Child of Krakatoa' has been growing apace and we can assume that at some time in the future it too will explode catastrophically.

On my 35th birthday I flew to Hong Kong to begin my annual leave. After checking into the Repulse Bay Hotel I joined a party given by the friend I'd made on the charter flight from London the previous December, Doreen. It was a pleasure now to be able to speak together rather than communicate by handwritten notes. The following day the news reports were dominated by the approach of typhoon Freda, bearing down on Hong Kong from the east, and I was intrigued and pleased shortly to witness my first major tropical cyclone. It hit the colony at midnight on the 17th June with ferocious winds and torrential rain that confined the population indoors and left the road across the Peak blocked by landslides and debris.

The next stop on my way back to England was not surprisingly, Bangkok, since I could not shake off my fondness for the chaotic warmth of the place. As usual, I stayed at the Oriental Hotel – still a hotel of modest size – and enjoyed being greeted by the staff as a regular guest. Calling on Noel at the office the next day, he told me that Ken Waller was in town and would like to have a chat with me. Ken was a senior BP man and had now taken over from David Robertson who had been posted elsewhere. Our paths had crossed from time to time and I had found him a friendly and kindly man. He was also staying at the Oriental Hotel and we met in the bar. What followed completely took me by surprise. He explained that he had been receiving reports that I was something of a "barrack-room lawyer," and said that for the sake of my career I should desist. Although I can no longer recall the details, what he was referring to was the support I had given to one of the geologists in my team in a wrangle he was having with the Company over some personnel matter. His comments upset me more than a little, since to back up one's staff on a matter with which one agreed seemed only fair and reasonable. We parted without rancour, although it had taken the wind from my sails and I went up to my room feeling distinctly crestfallen.

I changed planes at Calcutta and made a side-trip to Kathmandu where I stayed for five days, exploring the Kathmandu Valley on a hired bicycle. I found the people charming and was constantly stopping to drink tea and talk with them. My base was the Mount Makalu Hotel – good value at US$ 5.50 per night including breakfast. West across the Vishnumati River I cycled, to the hilltop temple of Swayambhu Nath with its pairs of painted eyes gazing over the valley; to the Deopatan temple complex where a cremation was taking place beside the river, a gruesome scene which I found disturbing – the corpse's limbs sagged from the flames only to be flipped back by attendant male relatives with long poles. Dry grass rather than firewood was the main fuel which somehow made the cremation the more pitiful, as this was a poor person's cremation I was told. I cycled to Patan, to Bodnath where young girls were weaving carpets. And back in the city a procession was taking place, and I watched King Mahendra returning to his palace after a trip to the Soviet Union and Afghanistan. The next day, while wandering in the bazaar at Kathmandu I fell into conversation with one of the stall-holders. He claimed to be a soothsayer as well as a fruit-seller and offered me an impromptu assessment of my future. I no longer remember whether he foresaw a happy love life or a prosperous career, but after running his fingers over my forehead, presumably to check for wrinkles, he pronounced in a strong Indian accent *"You will be having, Sir, a life which will be lasting to 84 years of your age"*. At the time I congratulated myself on my good prospects, but as that age approaches I begin to wish he'd given me a bit longer.

Arriving in London at dawn on the 28th June I made my way to Liverpool Street Station, and after visiting the Great Eastern Hotel's barber to make myself presentable, I took the train to my parents' house at Harold Wood.

Geologists back on home leave were required to visit Head Office in the City. Over the following days I had discussions with colleagues responsible for the Far East and attended the formal work-programme and budget meeting for which I had been preparing in Singapore. It was on one of these visits that I learned that in the following year I was to be posted to Melbourne as Chief Geologist for the Australian and Papua New Guinea operations. I would be taking over from David Jenkins, one of Exploration Division's high-fliers. At least, I thought, my reputation as a barrack-room lawyer doesn't seem yet to have damaged my career.

Relations with Anne were cool, and although she made clear that she preferred I did not come to Billericay, I was not as yet banned. It was difficult to have any normal and relaxed contact with Helen and Sally and it put me in a despondent mood. My parents saw that I was cast down and generously offered me their car to enable me to take a trip to Scotland. I drove their tiny Anglia north through Edinburgh and spent a week camping in remote glens and walking the hills, feeling my cares slipping away. One happy memory is of a morning on the shore of Loch Laggan. I awoke to hear the rain on the canvas above my head, and then the scrunch of footsteps approaching across the gravel. It was the Dutchman I'd exchanged a few words with the previous evening as he and his wife set up their caravan. He crouched at the door of my tent and called to me in his excellent English: "We wondered if maybe you'd like a cup of tea."

At Harold Wood again my mother's brother, Eric, invited me to join him on a yachting trip to France. I'd long since forgiven my Uncle Eric for getting me a spanking when he reported me to my father for swearing, and so was able to enjoy a sailing trip to Cherbourg and back in his Nicholson 32, with John my fellow crew member. At the pretty town of Saint Vaast La Hougue we tied up to the harbour wall and went ashore for a meal. A couple of hours later we emerged from the flower-bedecked *Hotel France et Fuchsias* and discovered we had seriously miscalculated the tide – or more likely neglected even to consider the tide. Eric's beautiful yacht *Sombra* was hanging high and dry on the harbour wall, its mast at a crazy angle and its mooring lines as taut as violin strings. There was nothing we could do but wait for the tide to come back in, but it was a painful lesson for Eric.

In August BP's medical department referred me to a Harley Street specialist to investigate my continued nagging bowel problems. I was shortly admitted to the Hospital for Tropical Diseases at St Pancras where I spent the last days of my leave giving stool samples and being dosed with Tetrachlorethylene for hookworms. This medication – once used as cleaning fluid before its poisonous nature was appreciated – was taken by mouth, and when my time came to swallow my dose it had an effect which none of the nursing staff foresaw. As I lay there on my bed, they saw that my breathing was becoming unusually shallow. When they tested my pulse they were unable to find one. The ward immediately swung into emergency mode; the ward sister saw my life ebbing away and shouted urgently for oxygen and heart stimulation. I recovered, of course, and ascribe my dramatic reaction to the unusually low blood pressure and low pulse rate that I'm told I normally have. If the medication had not been fatal for me it apparently was for my hookworms and I was not troubled again.

It was good to be back in Singapore again. Stuart and two Singaporean girlfriends from the Swimming Club met me at Paya Lebar Airport and drove me back to my flat on Dunearn Road. Goon appeared from her cubby hole next to the kitchen and broke into a smile and a *'Hello Masser.'* She had been told of my coming and had got in the inevitable pork chops and spinach for my dinner. Everything seemed as I had left it. The southerly monsoon continued to bring sunshine and occasional rain, and the thermometer on my balcony registered its usual 32 degrees as if locked at that temperature. I showered, changed into a sarong, and relaxed with a long cold drink.

One of Peter Wood's more annoying characteristics was to while away his working day by coming into everyone's office in turn, simply to chat. And so when he went on leave in September, output from the office increased significantly. I was asked to stand in as Acting Exploration Co-ordinator, a temporary promotion.

I continued making short trips around the region, including one to northern Thailand where I was able to see the very small oilfield and refinery that were operated by the Thai Defence Energy Department. It was the cool season and the samples of waxy crude oil I collected at the wellhead solidified in the chill air into something like shoe polish.

An equally memorable trip was the one I made to the Philippines. An independent Filipino company named Oriental Petroleum had been awarded offshore acreage to the west of the island of Palawan. Their Managing Director, Oscar Manuel, had called on us in Singapore and let us know that they were looking for a partner for the exploration venture. In late November I flew to Manila with Dave Thomas, our geophysicist, to investigate this new opportunity.

It was a fascinating trip. Oscar was a generous host and he and his geologist, Emmanuel Tammesis, entertained us in fine style. We stayed at the Manila Hilton and were introduced to the colourful night-life of the city. With their natural sense of rhythm and their Spanish heritage, talented Filipino musicians can be heard across the capitals of Southeast Asia and there are countless clubs in the Makati district of Manila where they perform. After the orderly world of Lee Kuan Yew's Singapore it was amusing to see signs at the entrance to each club, requesting that patrons kindly leave any firearms at the front desk.

After a geological excursion with Oscar to the northern town of Baguio, famous for its terraced rice-fields on the steep volcanic hillsides, we returned to the Manila Hilton in time to join a cocktail party laid on for the hotel's guests. One of our hostesses was a tall, attractive, girl who made the rounds of the guests ensuring that we were well looked after. We talked together and I learned that Mary Lou – for that was her name – was the General Manager's Personal Assistant and furthermore, was the reigning lady water-skiing champion of the Philippines. Before the party was over I had obtained her address and promised to write to her.

As for the real purpose of our visit, Dave and I found that seismic surveys over Oriental's Palawan offshore area had revealed the presence of reefs, buried beneath thousands of feet of sands and mudstones. Such reefs (or bioherms as they are correctly called) are ancient build-ups of coral or shell debris which, because they tend to retain some of their original porosity, are prolific oilfields in many parts of the world. In spite of the virtual absence of any oil production in the Philippines at that time we recommended to London that this was an opportunity worth pursuing. Our proposal was turned down in Head Office, but it is satisfying to reflect that oil discoveries were indeed made in those reefs in later years, and oil production from them became an important source of revenue for the country.

Christmas in Singapore brought rainy weather and a round of parties, but I was distracted by a flourishing correspondence which had developed with Mary Lou. I was keen to meet her again and we agreed that she would come to Singapore for a visit. Having obtained leave from the

Hilton she arrived late at night on New Year's Day 1972, and over the following week a romance blossomed. Mary Lou Noble was from a landowning family on the island of Negros. This is in the southern part of the Philippine archipelago which meant that her mother tongue was Visayan, not Tagalog, although she could speak both – as well as Spanish and perfect English, the latter with an attractive Latin American accent. She was tall and gorgeous and had a warm and earthy sense of humour. In the year and a half I'd been in Singapore I had made friends with a number of girls and some friendships had developed into romantic relationships, but there was something special about Mary Lou. As the day that I was to leave Singapore approached, I began planning how we could meet again.

Reink Lakeman was to replace me when I left for Melbourne and he arrived in Singapore in mid-January for an extended handover. A Dutchman, he had been a geologist with the Company far longer than I, and this greater seniority suggested to me that in due course he would move up from my position to replace Peter Wood as Exploration Coordinator. It was important for him to meet our many associates in Indonesia and I arranged a full itinerary of meetings. A brief jotting in my 1972 pocket diary says that on Saturday 29th January we were in Bandung where, of all things, we lunched on goldfish and gouramis at a restaurant called, appropriately, Ikan Mas. (Either the choice was limited that day or we were feeling particularly adventurous). The following day we found the time to climb Tankuban Perahu, the 1830 metre active volcano which overlooks the city, before returning to Djakarta on a train hauled by an ancient locomotive fired not by coal nor wood but, perhaps unsurprisingly, by oil.

Reink formally took over from me on 1st February. Peter Wood hosted a cocktail party for me at his fine old colonial-style home, and there was a dinner party for me in the Churchill Room at the Tanglin Club. The packers arrived to crate up my few belongings at my Goldhill Towers flat, and on 14th February I left the island state.

From the standpoint of my career my time in Singapore had been an important step up. I had managed a team of people for the first time, and, because of London's curious staffing of the exploration venture, I'd fulfilled most of the tasks of an Exploration Manager. I left behind what I considered to be a well-organised and well-run exploration office. On a personal level it had been a period of self-discovery. I had married so young that I had never experienced until now the life of a bachelor. The freedom of it was exhilarating, although throughout I had an underlying feeling of dissatisfaction, as though I was searching for something. As a place to live, Singapore was pleasant, green, orderly and easy. It was also unexciting, although less so than it became in later years.

Before I finally left Southeast Asia I wanted Reink to see something of the geology of Thailand. And so we flew together by way of Penang to Songkhla, the small and attractive town on the Gulf of Thailand. By this time Geoff Larminie – an old friend and colleague from the times we'd worked together in the UK and Alaska – had taken over as the Company representative in Bangkok, and he flew down to meet us at the Samila Hotel. During my Thailand survey I'd watched as fighting bulls trained on the beach in front of the hotel, but now they were nowhere to be seen. It was the start of the Chinese Year of the Rat – my birth year – and the Chinese community was celebrating with fire-crackers when the three of us got to the small town of Trang the following evening. The next day we continued working our way across the peninsula in our hired car, stopping repeatedly to examine the geological localities I'd become so familiar with. From Krabi the drive to Phuket winds through some of the most stunning scenery in the country, a landscape of jungle-covered limestone pinnacles which march into the Gulf of Phangna through a wide zone of mangrove swamp and meandering rivers. It was the last I'd see of Southern Thailand for many years.

I could have flown direct to Melbourne but I still had some unused leave due to me and so I opted, of course, for the scenic route – via the Philippines. I flew from Bangkok via Saigon and was met by Mary Lou at Manila airport, the terminal still black and charred following a recent fire. The next twelve days were a whirlwind of new places and experiences which Mary Lou had organized. We did what all tourists do in Luzon, and shot the Pagsanjan rapids in dug-out canoes; we drove north through Tarlac and Dagupan City to a rest-house at Lucap and snorkelled among the Hundred Islands of the Lingayen Gulf; and, most memorably, we visited Mary Lou's family on the island of Negros in the Visayas.

The overnight voyage from Manila North Harbour arrived in Bacolod at eight in the morning, but it was another nine hours in a rickety taxi before we reached the family home at Dumaguete City. The island of Negros was then, and probably is still, well off the main tourist routes and for the local people a European was an unfamiliar sight. Bumping through dusty villages along the way, children would wave and call out "Hi Joe!" assuming that I must be an American.

We were given a warm welcome by the Noble family. Mary Lou's father had trained as a dentist and had a practice in Dumaguete, but what I was keen to see was the family's sugar plantation. We drove there the next day, over more dusty roads and past the so-called 'central' – the factory which processes the cane into sugar and its pulpy residue, bagasse. Rolling acres of cane spread as far as the eye could see, some of it still green and looking like forests of bamboo, some burned black and being cut by hand before being loaded onto lorries. Armies of workers are needed for this dirty and grinding work and they were housed with their families in lines of shacks. I felt as if I had been transported back in time to an Eighteenth Century plantation in the southern USA. There was no doubt who was in charge – it was Mary Lou's brother, Sonny, who did the rounds with an automatic pistol jutting from his belt. Perhaps that should not have surprised me, as Mary Lou herself never drove alone through the streets of Manila without her pearl-handled revolver under the seat of her car. In all my travels I'd never seen a society so ripe for violent revolution as those plantation workers.

But nothing could detract from the pleasure of the last twelve days and it was difficult to say goodbye to Mary Lou. Already I was hatching plans for her to join me later in Melbourne. The next day, 3rd March 1972, I arrived at Tullamarine Airport.

Chapter Twelve

David Jenkins met me and drove me to the Travelodge Hotel, conveniently across the road from one of south Melbourne's main landmarks, the twelve-storey, curved-fronted BP House.

My first week was taken up with luncheons which David had arranged for me to meet my counterparts in the other Melbourne-based oil companies, with evening buffet dinners to meet my colleagues, and with hunting for a place to live. I settled on a flat on the third floor of a modern block on Beaconsfield Parade, the seaside promenade in the Middle Park district. My flat was on the side of the building, and I had to strain my neck out of the window to get a view of Port Philip Bay, but it was comfortably furnished and had the advantage of being only a half-hour's walk through the park to the office.

I recall that I was not as excited nor as happy to be in Australia as I had expected. I had only seen glimpses of the country before, on my way to and returning from New Zealand, and perhaps I imagined that I would immediately enjoy being there. I daresay I fell into the same trap as many immigrants from the UK, the expectation that because Australia was settled by the British it would be like Britain – but Britain with a warmer climate. The reality is, of course, that nearly two hundred years had passed since those first settlements, and the country has developed in a direction of its own. Certainly, there is much that a British immigrant finds familiar, but somehow that only makes the differences jar all the more. A lot of what I saw and felt in those first weeks depressed and disillusioned me.

But gradually my prejudices fell away. I began to find the suburban streets of Victorian cottages were charming; I became absorbed in the unique flora and flora and spent my lunch hours in the beautiful Botanical Gardens close to the office; and perhaps most important, I began to build a circle of good friends among the Melbournites.

On the day I took possession of the flat, David and I began a series of trips around the continent for him to introduce me to others in the oil exploration business that I'd be working with. In Perth we stayed at the Parmelia (for David had no doubts about his entitlement to stay at the best hotels) and I spent some time listening to presentations by the management team of Burmah Oil, the operator of our important joint venture on the Northwest Shelf. A couple of days later we were in Canberra meeting the government's Bureau of Mineral Resources (BMR) people, then to Sydney and up to Brisbane to meet the management of Texaco who were our partner in one of our exploration blocks in Papua New Guinea, or PNG as we generally called it. The next stop on this whirlwind trip was Port Moresby, the capital of this still-Australian territory.

I was surprised to learn that BP first became interested in the territory some three decades earlier when it carried out an exploration programme on behalf of the British and Australian governments. In the interim the Company had gone through cycles of enthusiasm for the area, encouraged by the numerous natural seepages of oil and the immense size of many of the anticlines – anticlines which reminded BP's older geologists of the prolifically oil-bearing anticlines of Persia. Because PNG was not yet independent it was subject to the same very attractive 'subsidy' rules as mainland Australia. Under those rules the Australian Federal Government contributed toward the cost of approved oil exploration operations, including drilling, and there is no doubt that this was a spur to the oil companies. Arriving in Australia in 1972 I found that BP was active in a number of PNG licence areas, notwithstanding the two dry boreholes it had drilled in the Southern Highlands in the year or two before I took up my new post.

From the hot and humid, sleepy, capital we flew by Fokker Friendship to the principal town of the Southern Highlands, Mount Hagen, and then by light plane to the village of Erave which David Balchin, the leader of our current geological survey, had made his base. This was wild and remote country and no sooner had our plane come to a standstill than a group of Highlanders gathered around it. The men were naked apart from their grass skirts, and around their necks hung large ornaments of mother-of-pearl. Their faces were painted, their woolly hair was decorated with flowers, and through the nose they wore a boar's tusk – or in the case of the more worldly ones, a ball-point pen.

This was no show put on for tourists – there were no tourists – but was their everyday dress. Only later, in even more remote Highland villages, did I come across men dressed in the item for which PNG has become famous, the penis gourd.

It was cool and damp when we arrived at Erave and that night the heavens opened and a tropical downpour flooded the floors of our tents. But the next day the clouds had lifted enough for us to carry out an airborne geological excursion of the Highlands, stopping at Mendi to meet the District Commissioner and at Kiunga where our seismic survey was underway and where we were able to refuel our Baron aircraft. In spite of the complete cover of rainforest the elements of the geology were easy to see: the huge limestone whalebacks of the anticlinal folds that seemed to hold such oil promise (but which so far had proved so disappointing), pierced here and there by the cones of active or dormant volcanoes. Nowhere had I seen such rugged terrain, such immense jungle-covered escarpments, with waterfalls seeming to pour out of the clouds.

Back in Port Moresby, at the Gateway Hotel, David introduced me to a trio of Japanese who, he explained, were with a company called Oceania. (I should explain here that whereas BP had licences to explore a number of areas in its own name, other areas were licenced to Australasian Petroleum Company (APC), a joint venture company in which BP as well as Mobil and a small Sydney-based company called Oil Search were shareholders. We were the designated operator on behalf of the group, and these APC licence areas were therefore also in my bailiwick). I learned that Oceania was negotiating to farm in and drill some wells on one of APC's licence areas, and over the following years I had a lot of dealings with the Japanese and grew to like them: Naganuma, the Managing Director who grew orchids in his spare time and was based in Tokyo, Yoshi Hayashida their urbane Exploration Manager, and 'Nobby' Miyazaki their geologist. I recall an occasion some months later when we were lunching together in Melbourne. Yoshi was explaining some geological idea of his and was attempting to sketch it on a paper napkin. The pen he'd taken from his pocket wouldn't work and in exasperation he threw it aside and expostulated "Oh, these bloody Japanese pens!" I hadn't previously experienced a Japanese sense of humour and had doubted whether they had one. Many years later, and married to a Japanese, it's hard to recall I ever harboured such doubts.

On 27th March I formally took over from David as Chief Geologist. This was a bigger job than I had in Singapore, both in terms of the greater number of geologists in my team and, more particularly, in view of the much more active exploration programme we were involved in. We had a continuous drilling programme under way on the Northwest

Shelf where several major gas discoveries were at the appraisal stage; joint ventures offshore Perth and in the Otway Basin west of Melbourne; geological surveys which were intended to lead to further drilling in Papua New Guinea; and our office was reviewing the other sedimentary basins of Australia with a view possibly to obtaining acreage in them. The team numbered eight expatriate geologists which I considered just sufficient for our workload.

Someone once described Australia as suffering 'the tyranny of distance.' Although they were referring to the country's remoteness, they might equally have had in mind the distances within the country itself. A car is a near-necessity and so I was pleased when, a month after I arrived in Australia, the ship carrying the Mercedes I'd ordered from Germany reached Melbourne docks.

With this new mobility a priority was to call on my good friend from my days in Bangkok, Rod Hood. He and his charming new Thai wife, Rudi, lived in the seaside suburb of Sandringham. Also, through an introduction from Singapore friends, I met the Pannell family. Richard was a lecturer in English literature at Monash University and had met Sujatha, who later became his wife, when he had been on the staff of the University of Singapore. I liked the Pannells immediately and they became, and have remained, good friends. Their house in the cool and misty Dandenong Hills became, like the Hoods', another second home during my stay in Melbourne.

There was a lively social life too among the exploration people who were my colleagues on the eleventh floor of BP House. Some of them I'd worked with before, including my boss, Ken Roberts the General Manager, an Australian now happy to be working back on his native heath and who I'd last seen at Eakring fourteen years earlier. It was a pleasure to be working for him again.

Having given up her job with the Manila Hilton, Mary Lou came to live with me at my flat on Beaconsfield Parade. This was the first time she had lived outside Asia, but she soon settled into her new life. We had many enjoyable times together and her sense of fun made her popular among my friends, particularly Rudi with whom she developed a lifelong friendship. I remember how she would call out with delight when she spotted apples growing in someone's garden – she had known them only from greengrocers' windows before. And snow was another new experience she was able to enjoy when we drove to Falls Creek in the Great Dividing Range for a weekend of skiing with the Hoods and with David and Gloria Balchin. Public holidays gave us frequent opportunities to explore beyond the Melbourne region, including a trip on the Queen's Birthday long weekend to the range of sandstone mountains called the Grampians, and on further west to see some of South Australia.

But my work meant that I was often away from home. In June I returned to London for a week of meetings on staff and work-programme matters at Head Office, returning via Singapore to see old friends and colleagues. I had only been back a few days when I had to leave again for Papua New Guinea, this time to see how Martin White was progressing with the Lagaip River geological survey.

It seemed perpetually bad weather in the Highlands. Each morning we would be dropped by helicopter at a freshly-cut jungle clearing, from where we would wade up river beds and follow our bare-foot Highland field assistants as they hacked a way through the rainforest. Often we would be halted temporarily by sheer-sided karstic limestone pinnacles and cliffs, but our assistants were skilled at building makeshift ladders from jungle timber and vines. At nightfall we would cut another clearing and be picked up by helicopter to return to an uncomfortable and rain-drenched camp . It was an awful environment for a geologist to have to work in, and yet most of my colleagues, if not actually enjoying the life, accepted it without a murmur. Martin himself seemed particularly at ease in the jungle, his strong and wiry frame nimbly negotiating fallen trees, waterfalls and gullies. Not for nothing had he gained the nick-name from his fellow workers "Speed-of-Light-White."

After nearly six months together it was becoming increasingly clear that Mary Lou and I would have to go our separate ways. I was not yet ready, I realised, to be in a committed relationship with one woman. With sadness I dropped Mary Lou at the airport, and she left for Honduras where she planned to stay for a while with friends. We kept in touch and remained friends, and a few years later she met and married an American, settled in the USA and raised a family. This was probably what she had dreamed of, as she always used to refer to herself laughingly as "a little brown-skinned American."

Trips to Perth became a routine occurrence, to attend meetings with Burmah Oil and our other joint-venture partners. We would stay for a couple of nights, enjoy some good meals hosted by Burmah, and sometimes even have time for a swim at Cottesloe or one of the other Perth beaches. I had a high regard for the work that the Burmah team was doing on our Northwest Shelf acreage and the pace of activity was lively, driven by the constant need to devise technically-sound exploration targets to ensure the contracted drilling rigs were never idle. And as well as having exploration prospects to discuss there were appraisal-drilling programmes to agree, to bring the Goodwyn and North Rankin gas discoveries to the point they could be declared commercially viable and work could begin on laying a pipeline to bring the gas ashore. Our partners included Shell and, since they too were based in Melbourne, we would sometimes fly across to Perth together. My opposite number as Chief Geologist was one Mark Moody-Stewart. I got to know him well and, although he had the somewhat buttoned-up manner typical of Shell men, I had to admire the obvious preparation he did for these meetings, and was impressed by the cogency of his arguments to persuade Burmah to the Shell point of view on any matter. His abilities obviously did not go unnoticed in his own company, and he went on to become Shell's Chairman, for which he received a knighthood.

In the southern spring of 1972, one of these meetings was scheduled to take place over a Thursday and Friday. Although our party flew to Perth in the usual way, I decided with another colleague, Tullis Sutherland, to return to Melbourne by train. The idea of travelling the longest straight stretch of railway in the world was an obvious attraction. After a convivial dinner we left the others to await their midnight flight back to Melbourne, and Tullis and I departed for the railway station. The *'Indian Pacific'* left Perth at 9.30 p.m. on Friday night, and we climbed aboard and were shown to our sleeping compartments. Before turning in we met in the comfortable club car for a nightcap, relaxing in armchairs while a fellow-passenger picked out tunes on an old upright piano. My sleeper cabin was spacious but I was surprised that it was impossible to turn off, or even turn down, the broadcast announcements by the train's conductor. I awoke the next morning to his cheery greeting and found we were in Kalgoorlie.

As the train rolled eastward the scenery became redder and more sunbaked, and the trees thinned until there were none – we were crossing the Nullarbor Plain. In spite of the aridity, occasional groups of kangaroos could be seen bounding into the distance, flocks of galahs and sulphur-crested cockatoos took flight, and emus would turn their heads to gaze at us as we rumbled by. At midday on Sunday we arrived at Port Pirie. The main line continues east from there to Sydney, but to get to Melbourne we would have to go via Adelaide and catch the *'Overland'*. The highlight of the whole trip was without doubt that leg from Port Pirie to Adelaide. Chatting to the engine driver before the train was due to leave, he invited me up onto the footplate and I spent the three and a half hour journey to Adelaide sitting beside him, watching the way ahead and learning the ins and outs of being a train driver. We arrived at Melbourne on Monday morning and I was back in my office only a half-hour late, a schoolboy dream realised.

My circle of friends outside the office was widening and I was now thoroughly enjoying life in Australia. The misgivings I had in the days after my arrival were now quite forgotten. There were visits to the theatre, and I remember a jazz concert by Dave Brubeck, and a day at the Melbourne Cup (at which, unfortunately, I failed to spot the winner, Piping Lane). Often I would be invited to dinner parties at friends' homes and although I returned their hospitality my cooking was generally of the basic kind. I remember a happy day's riding with Sue from the office at her home on the Mornington Peninsula, a day when – to my own surprise – I established my manliness by brushing from her bosom a giant spider which had found its way onto her while she gathered hay for the horses.

And as the summer advanced I would swim off the beach in front of the flat. There were days when the weather was hot, and a dry searing wind took temperatures up to 40°C. But depressions tracking through the Tasman Sea could produce very sudden changes. In the space of a half-hour the wind might back to a southerly, the temperature would drop to below 20°C and bathers would bundle up their belongings and scamper off the beach as though there'd been a tsunami warning.

Taking advantage of the long days of summer, a number of us would meet after work to play tennis in the park on Toorak Road, finishing the evening at the nearby Fawkner Arms. Squash too was popular, and although I'd improved little since first picking up a racket in California, it was a game I still enjoyed a lot. By contrast, some of my colleagues were expert players.

One of my BP colleagues, an Englishman I shall call Andy – not his real name – was a keen and very competent squash player. He had heard me talk about Rod Hood and the fact that Rod had been the Thailand champion, and Andy was keen to meet him on the squash court. I duly put them in touch and they agreed to play at a nearby club during the lunch hour one day the following week. The match was, according to Andy, an enjoyable and a close-fought one and he looked forward to another contest some time. The following weekend, relaxing at the Hoods' home, I asked Rod how the match had gone. It had gone well, he said, and Andy was certainly a good player. Then he went on in his colourful Australian accent:

"But jeez, Mike" he said, "I know you Poms have a reputation for being a bit shy at the sight of soap and water, but I never thought I'd get to see the likes of your cobber, Andy! After playing for 45 minutes we were both sweating as if we'd been in a Turkish bath, when what do you think happened?"

"I've no idea" I responded, "what did happen?"

Rod continued, in tones of rising incredulity:

"As I stripped off to get under the shower, what does your Pommy chum Andy do? He pulls on his bloody shirt over his sticky body, gets into his strides, puts on his jacket, and with sweat still running down his red face, climbs into his car and heads off back to the office!"

Laughing at Rod's account of the unwashed Pom, I tried to assure him that Andy was probably in a hurry for an important appointment at the office, although privately I couldn't but agree that the Brits' reputation is not entirely undeserved. I thought back to the 'Sunday night is bath night' regime of my own childhood and was thankful that my years among the scrupulously clean people of Southeast Asia had finally removed any lingering traces of it.

Australia used not to have a high reputation for its restaurants, but by the 'seventies there were many good ones. I developed a new familiarity with wine, and a taste for the Cabernet Sauvignons and the Pinot Noirs of Australia's vineyards. Often we would go to a bistro with a 'BYO' sign on the door, something I'd not come across elsewhere but was, I thought, a useful Australian invention.

One day I was in downtown Melbourne, and since it was lunch time I decided I'd drop into a restaurant for a quick meal. The place I chose seemed pleasant enough and had an ineresting menu displayed in the window. A waitress ushered me to a table and I settled myself and examined the menu. The fish looked good and so when she returned a few minutes later and asked me what I'd like, I put down the newspaper I was reading and replied I'd have the whiting. I carried on reading my newspaper. Twenty minutes passed and my waitress by this time had served other customers their lunch, even customers who had come in after I had. I couldn't understand why she was neglecting me, and so tried to catch her eye. By now I was hungry and irritated that the service was so poor. Eventually she saw me waving my newspaper at her and she came over.

"Yees?" she said, in her strong local accent.

"I've been sitting here over half an hour now" I expostulated, "for Heaven's sake, when are you going to bring my lunch?"

She seemed genuinely puzzled.

"What lunch?" she asked.

"The whiting I ordered" I replied with a resigned tone.

"Oh croiky!" she blurted, "I'm reelly sorry, mate! I thought yer sid yer were witing!"

Before Christmas I made another trip to Papua New Guinea where, again, Martin was leading a geological survey, this time over an upland limestone platform which was the surface expression of a pair of upfaulted geological structures called the Darai and Kanau anticlines. This was preparatory to our Japanese friends in Oceania Petroleum drilling a pair of exploration wells a couple of years later. It was unremitting karstic terrain, and carrying out traverses to collect rock samples and measure the angle of dip of the limestone beds was slow and tiring work. It was a relief when the Cessna arrived some days later and took me to Mount Hagen. In a letter home written from the Hagen Park Motel on 2nd December 1972 I wrote:

> *"I arrived by light plane from the camp at Bosavi at 10.00 this morning, just as the Ansett flight was leaving for Port Moresby and the south. And so I'm stuck here till tomorrow when I'll be able to get back to Melbourne. It's been a good spell in the field, but in an incredibly rough terrain of jungle-covered limestone pinnacles. To travel a mile horizontally means about three in all, with the ups and downs. The first day out traversing nearly killed me, and on the last climb to the helipad I'd need a rest every ten minutes to allow my pulse to drop back from a rate which I measured at 135 – just by hearing it thumping, not by feeling my wrist. But by yesterday I was much fitter and a five-hour traverse couldn't faze me. Instead of having to collapse to rest and have the leeches get me, I could rest upright so that the blighters couldn't get any closer than my jungle boots.*
>
> *There's a lot of camaraderie on a survey like this and the evenings are a lot of fun, when everyone is cleaned up and replete from a good dinner, usually with a bottle of wine.*
>
> *This afternoon I witnessed a bit of a tribal skirmish here in town. I was out for a stroll to freshen up after a post-lunch snooze, and heard a lot of shouting, as if it were a football match. Natives, many in their tribal grass skirts and headgear, were swarming toward the source of the din, and so out of curiosity I wandered in the same direction. Then the crowd seethed toward me and I saw that a running stone-throwing battle was going on. It was in earnest too. Pretty soon the police arrived and started to clear away the onlookers. I asked the police chief, a European, what was the trouble and he very rudely replied "Mind your own business and go away if you don't want to get hurt." He was in quite a state of upset. A month or so ago a number of natives were killed in a tribal battle like this, so I guess that was the cause of his agitation.*
>
> *I hope my potted plants will still be alive when I get back home. Perhaps I should have asked the woman upstairs to look after them. Funny! The other day as I was sweating through the jungle I thought that one could make a fortune from the plants there: Philodendrons, Monstera and all kinds of others grow in lush*

profusion, forming a matted tangle. On the sharp pinnacle limestone you're often walking or climbing over a web of this stuff like a spider, often ten feet above the ground.

Another way to make a fortune would be to build an hotel at Beaver Falls, on the Kikori River. These are a magnificent sight, possibly as spectacular as many of the world-renowned falls, but in the depths of the jungle and so unknown to anyone outside Papua New Guinea. To fly in a chopper down at the bottom of the gorge with a wall of water on three sides is a memorable experience.

Would you believe that sitting here in the hotel lounge a half hour ago I felt an itch and found a tick burying itself in my arm. I dug him out but think I'll go and have a good bath now ..."

Back in the relatively modern world of Port Moresby the next day, I caught my plane for Brisbane. But even this turned out to be not without incident. We had been flying for an hour or two and were somewhere over the Coral Sea when the captain announced that we were turning back to Port Moresby having been informed of a possible bomb on board. Remembering that only three months previously the Israeli Olympic team had been massacred in Munich, and later that month, September, bombs had exploded in Sydney, we passengers on that TAA flight took the news seriously. Back in Moresby the plane was parked at the far end of the runway and searched while we were bussed to the terminal to wait. After several hours we took off again, but it was midnight by the time I got home that night.

While I'd been away the Federal General Election had taken place and Gough Whitlam had become the new Prime Minister, leading the first Labor government for over two decades. Among my left-leaning friends there was rejoicing although, with thoughts of Harold Wilson's recent six-year reign in Britain, I was less sure. The Australian trade unions already had enough power, I thought, and strikes were an all-too-frequent feature of life.

One of the results of the change of government did have a direct affect on me and my colleagues. Workers in the oil industry were awarded a nine-day fortnight, which meant that we became entitled to one extra day off every two weeks. This was, of course, very pleasant for those whose turn it was to take the day off, but administratively it was a headache. Each company was expected to arrange rosters for its employees but no mechanism existed for co-ordinating the scheme across the industry as a whole. For us exploration people, most of whose operations were carried out in joint ventures, organizing meetings or even managing to speak to a particular person in another company became frustratingly difficult. On, say, Thursday I'd try to call so-and-so only to be told that it was his day off. The next day he'd call me back, but that day, Friday, happened to be my day off. And so it went on. It was a scheme of loony-left barminess.

The hot and dry summer weather continued and was causing a lot of concern to everyone. In a letter home in late December 1973 I wrote:

> *"Melbourne has been having occasional days with temperatures above 100 degrees (Fahrenheit). It's a fierce heat and the vegetation becomes tinder dry, and so on such days a 'Total Fire Ban' is declared throughout Victoria. In spite of this, over a hundred forest and grass fires started in one day recently, a lot of them started I'd guess by careless smokers, but most of them by spontaneous combustion of the eucalyptus vapour in the air. December has been the driest one on record, and coming after a particularly dry year the drought is getting very severe. Sprinklers are no longer allowed, hoses can't be used for washing cars, and stiffer restrictions are being talked of. Lawns throughout Melbourne's parks and gardens are now completely brown, and trenches have been dug around trees to conserve what little water is allowed for them. Really heavy rain is desperately needed to fill the reservoirs, but there's no indication that rain is likely ...."*

A month later I was able to write:

> *" ...the drought is petering out and on Tuesday we had almost an inch of rain which brought a green blush to the lawns. There's been no announcement yet as to how much the reservoirs have managed to catch and so restrictions are still in force. Meanwhile, over in Western Australia there have been a couple of typhoons, Kerry and Maud, which caused quite a bit of concern and damage to our offshore rigs."*

I was still in Australia two years later, when Typhoon Tracy hit the Northern Territory on Christmas Eve 1974, destroying seventy percent of its buildings and killing 65 people.

The unbroken dry weather was perfect for partying. I remember that on a fine warm Christmas Eve a friend, Gael, and I packed a hamper and a bottle of wine and while seated on the baked grass sang carols by candlelight at the Myer Music Bowl. Then on Christmas morning I was invited to Richard and Sujatha Pannell's home in the Dandenongs, and that afternoon we gathered at the Hoods' place and lazed by the pool. New Year's Eve came and the Pannells gave another of their memorable parties which went on till the early hours of 1973; to this day they won't let me forget my cruelty in getting them from their bed just a few hours later, and dragging them off for a scorching-hot day of picnic races at Hanging Rock, a picturesque country racecourse an hour north of Melbourne made famous a year or so later by one of Australia's early film successes, "Picnic at Hanging Rock."

I came to delight in the "great Australian outdoors" and whenever the chance arose I'd drive into the countryside, alone or with friends. By now I'd bought a small, two-man tent (or, as I described it to my parents, a one-man-one-woman tent) and it saw a lot of use. At the height of summer a group of us, including Melody from the flat upstairs, headed for Wilson's Promontory, a granite peninsula which juts south into the Tasman Sea. For several days we trekked around its perimeter, over scrub-covered headlands and along wide empty beaches, stopping to swim when the heat got too much and camping at night around a driftwood camp fire. And I remember another trip in the old gold-mining district of eastern Victoria where we camped on the bank of the Snowy River, and where I successfully panned three tiny grains of gold from the river gravel.

In the winter, camping was less attractive but my journeying around the state continued. I described one winter trip in a letter home:

> *"I spent the long weekend in western Victoria, staying on a friend's farm on Friday night and in a delightful little cabin in the bush on the banks of the Glenelg River for the next couple of nights. The weather was mostly cold and rainy but on Sunday night it was clear and starry and we were able to have a barbecue. A good log fire kept the cold at bay and it was fun to watch the possums come down from the gum trees to eat bread and apples we threw for them. They're the size of large cats and their heads are kangaroo-like, with pointed ears and large round eyes. At times we had as many as five within a few yards of us. During the day kookaburras would sit outside the cabin and fly down for scraps we threw for them."*

But it was not enough – I wanted to own a bit of the Australian bush. The Melbourne Age carried a weekly property section and I scanned it with maps spread out beside me, marking the bush blocks that attracted me and assessing their topography and accessibility. I settled on a two-acre block west of the small town of Healesville, and in due course it became mine. It occupied a south-facing hillside and was completely covered in eucalyptus forest. I loved the place. Over the following weeks and months I would drive there, bump up the steep rocky track and park my car, unload my tools and spend the rest of the day felling trees, cutting paths, and excavating by hand a small area of level ground. Sometimes Ravi, the Pannells' young son and a keen backwoodsman, would keep me company. At last I had levelled a grassy camp site where I could pitch a tent, and had felled enough trees to open a view of the rolling farmland of the Yarra Valley. Often I spent an entire weekend there, waking to the dawn chorus of currawongs and bellbirds, cooking over a log fire and delighting as the sun went down and the small marsupial honey-gliders emerged. And there were larger animals too, shy wallabies and once a five-feet long goanna which climbed out of sight up a large gum tree when I disturbed it. Before long I had identified most of the animals, birds and plants which I found there. The two most common trees were the stringybark and the bundy, both eucalypts, and I named this little corner of Australia, 'Bundy Hill.'

At around that time, the oil discoveries being made by BP in the UK North Sea and in Alaska brought the need for large capital expenditures to bring them into production. Inevitably they were given priority in the BP group, which meant that in other areas some belt-tightening was necessary. My job as Chief Geologist in Australia was not just to supervise the work of my team, but to map out a strategy for petroleum exploration in my region. Against that background of financial stringency in the Company, and with the termination that year of the Australian Government's petroleum exploration subsidy scheme, it was hard to remain up-beat about further work in the Southern Highlands of Papua New Guinea. The last two boreholes, drilled shortly before I took over in Melbourne, had been disappointing and I was coming round to the idea that the large anticlines – so clearly visible at the surface and seemingly so simple – concealed at depth strata which were complexly folded and, with the seismic technology then available, were likely to be impossible to elucidate. I spent much of my time at the office reviewing the geological findings from the boreholes and integrating the data gathered by our field surveys. Our most recent hole, Mananda No 1, drilled shortly before I'd arrived in Australia and on which so much hope had been pinned, was dry. I learned that compromises over its location had been made in order to be closer to a supply of surface water for drilling, but in doing so the well had penetrated the thrust fault which outcrops along the south flank of the anticline, missing altogether the target sandstone. A second well down the north flank of the Mananda Anticline might be justified, I concluded, but it would be a high-risk strategy.

I argued instead that we should concentrate our efforts further west in PNG, and that as a first step we should carry out a seismic survey over the Lavani Anticline. This is a breached anticline where the limestone carapace has been removed by erosion, revealing its deeper levels, and where there was good evidence (from outcropping granite along the trend of the anticline, to the west) that the fold was less structurally complex than, for example, Mananda. The eroded core of this anticline forms a broad valley several miles wide and ringed by high cliffs which are the escarpments of the eroded limestone carapace. This was the valley which, twenty years previously, had been 'discovered' by John Zehnder and, as I've related earlier, his discovery of a hitherto unknown tribe caused a flurry of excitement in the world's press. With helicopters it would now be possible to carry out a seismic survey in the valley to unravel the deeper structure of the anticline and, in particular, to establish whether the anticlinal 'closure' (that is, an outward dip of the strata) existed along the axis of the anticline as well as down its flanks.

Head Office accepted the seismic survey recommendation and Glyn Thomas, our Chief Geophysicist in Melbourne, swung into action with its planning. Meanwile two more geological surveys were under way over and around the Lavani Anticline. One of these was led by Angus Findlay and I spent some time with him in the field, a trip bedevilled by recurring problems with our helicopter which were only finally rectified by fitting it in the field with a replacement engine. A quiet, slim man with an attractive courtesy and gentleness, Angus gave the impression that in the jungle he would have the greatest difficulty coping with the heat, the humidity, the strenuous field routine and the discomfort of camp life. In fact the opposite was the case. This Harrovian who seemed such an unlikely explorer had the same grit and endurance as one sees in the best British explorers, and a comparison with Wilfred Thesiger would not be out of place. We have remained good friends to this day.

As time passed I saw the size of my geological team depleted as, one by one, their tours of duty in Australia came to an end and they were posted elsewhere. Soon their numbers had dwindled to five and I argued with London that this was insufficient for the workload they expected me to carry out. But my requests for additional staff were met with the response that, if I needed more, then I would have to recruit them locally. And so this I duly did. I found two good men to join us, one of them a British geologist with the West Australian Geological Survey named John Nicholson. John effortlessly made the change from mapping dam sites in the Pilbara to exploring for petroleum. Later, when I had moved on from Australia, I persuaded the Company to take him on to its worldwide establishment and was pleased when he was then able to join my team on my next posting, in Scotland.

The 1973 work-programme meetings in London were held in July and, again, I was required to present to management – Jack Birks and old friends from Palos Verdes, Alwyn Thomas and Jim Spence – the results of our year's exploration and our proposals for the year ahead. Although First Class travel within Australia had recently been restricted to more senior staff, on international travel I was still able to enjoy this perk. I chose to break my journey at Bali and spent several days based at the delightful Tandjung Sari on Sanur Beach. On a hired bike I explored Denpasar and the surrounding countryside and was completely charmed by the people and by their culture. It seemed they were incapable of making or doing anything without it being exquisitely beautiful. In the evenings it was enough to sit on the terrace of my tiny thatched cottage at the Tandjung Sari and listen to the gamelan orchestra playing by candlelight in a bower of tropical flowers and vines.

I spent as much of the ten-day business trip as possible with my parents and with Helen and Sally. Relations with Anne were difficult although she did agree to join me and the girls in an evening at the theatre when we saw Agatha Christie's 'The Mousetrap' at the Ambassadors. And at the girls' school Open Day I watched Helen take part in a swimming display, and in the evenings looked on as Helen and Sally rode their pony, Griffy.

On these business trips back to London I generally stayed at the Inn-on-the-Park near Hyde Park Corner. (It was an extravagance, but to paraphrase my colleague David Jenkins: "if the Company wants the best geologists, why should they get them on the cheap?") It was widely believed that the eccentric American millionaire Howard Hughes had taken over the top-floor penthouse and was living the life of a recluse there. I found this to be true one day when I entered the lift in the hotel lobby and absentmindedly pressed the wrong button. As the doors opened and I stepped out, a couple of burly henchmen got up from their seats and barred my way.

Perhaps it was my ownership of Bundy Hill, or simply that I was growing increasingly attached to the country, but I felt sensitive to its shortcomings and wanted to do something about them. I wrote to the newspapers about the rubbish that motorists threw from their windows, and was pleased when a hard-hitting campaign was started which showed a pig and its offspring rooting among beer cans and plastic bags, with the words "Only pigs litter!" It seemed to be effective, and over the years that I lived in Australia there was a noticeable improvement. Another issue which exercised me was the bizarre "give way to your right" rule of the road. Under this curiously Australian system you needed to be constantly alert to the danger of another vehicle coming quite legally out of a minor side-road directly ahead of you, even though you may be driving along an obviously major road. I could contain myself no longer, and on the 20th September 1973 my letter to the Melbourne Age was published. I kept a copy, and it ran:

> *"SIR, – The roads, or at least the intersections, of Australia are not paved with gold but with broken glass – from the windscreens of vehicles which have collided.*
> *Your editorial ('The Age', 13/9) focuses attention again on the utter confusion surrounding the right-of-way at road intersections, and the ineptness of the authorities in rectifying this.*
> *As someone who over the past 16 years has lived and driven in eight foreign countries, from Abu Dhabi to Alaska, may I say that in only one of these, Australia, have I noted confusion at intersections.*
> *The golden rule which applies elsewhere is that you can assume you are on a major (priority) road with right-of-way unless there is some form of sign telling you otherwise. It is then unnecessary to have priority road or priority intersection signs; it is the minor or non-priority road which has a sign saying "Stop" or "Give Way" on entering the major road.*
> *This makes for a smooth flow of traffic and eliminates confusion. Isn't that what we want?*
> *I cannot believe that Australians are so basically different from everyone else that this system would not work here."*

Change did not happen overnight, but some years later on a visit to Australia I was glad to see that commonsense had at last prevailed.

Soon after finding my feet in Australia I joined the Australian Petroleum Exploration Association and in 1973 was elected President of its Victoria chapter. As well as lobbying the government on matters such as taxation and the regulatory system, we had a programme of monthly meetings, and as President it largely fell on me to organise the guest speakers. At one luncheon meeting we had John Gorton, who until 1971 had been the Liberal Party Prime Minister. Although a "pleasant enough cove" as I referred to him in a letter home, I found his address and our conversation over lunch disappointing. Another was Don Chipp, a Liberal Party MP and until the 1972 change of government the Minister of Customs, who was altogether more interesting. A craggy-faced man with a relaxed air and an interest in Southeast Asia, I had met him socially on a couple of occasions through a friend named Virginia who had been his P.A. I remember a small dinner party at Virginia's home in Melbourne. Don was there with his then wife, Monica, and after the meal was over we sat around the table drinking port and talking until the small hours. He was one of those rare things, a likeable politician, and as well as advocating more liberal censorship and greater protection of the environment he confided to us that he was already planning to leave the Liberals and found a new party, the Australian Democrats. I had long since departed Australia when I read in 1977 that he had achieved this ambition and gained two seats in the Senate.

If that account of my professional life gives the impression that I was by now a well-rounded oilman, that is only partly true. In October the Yom Kippur War took place and the Arab embargo on exports to USA and other Western countries caused the price of crude oil to quadruple. Of course, the economics of our exploration were significantly affected and I daresay I should have taken more interest in this commercial side of the business, but the truth is that for me it was still geology that occupied the centre of my working life.

Letters from home made frequent references to the oil crisis and the shortages affecting Europe. But I was more concerned about the references in my parents' letters to their health. My mother had been having pains in her chest and it was diagnosed as angina. She should, of course, have taken life easier but that was not in her nature, and she continued to do her shopping and her housework at twice the pace of a normal woman of her age. And my father, who had referred to "problems with his waterworks" for several years, now found that he needed to have his prostate gland removed and receive regular painful diathermy and drug treatment for what he called warts in his bladder. In other letters he commented on disabling periods of dizziness. These were worrying developments.

In November, Angus Findlay and I made a trip to Perth, and after a meeting with Burmah Oil (which was our main reason for the visit) we joined a geological excursion led by the West Australia Government geologist, Phil Playford. It included a visit to the Permian rocks around Irwin River where certain conglomerate beds contained striated boulders and other features which showed beyond much doubt that around 280 million years ago Australia was undergoing an ice age. Although of no direct relevance to the search for petroleum, it was of particular interest to me in view of my own work on the fit of Southeast Asia into the ancient super-continent of Gondwanaland. When we reached Cockleshell Gully the excursion came to an end and the participants dispersed, but Angus and I decided to extend our stay in WA by making a journey further north. In a hired Landrover we zigzagged our way about 300 miles up the coast to Geraldton and beyond to the sandstone cliffs at Kalbarri, studying the geology as we went and having encounters with snakes and the bane of the dry Australian outback, myriad flies. I recall that at one point on this mini-expedition, as we bumped along a rough track to look at some rock outcrops, a strand of barbed wire blocked our way. Hanging from it was a sign cut out of an old rusting oil drum, and on it was scrawled in hand-painted letters "Private. Piss Off!" Only in Australia, we laughed, would you come across a sign like that.

That same spring I received a call one day from the boss of Interstate Oil, a subsidiary of the large mining company CRA. It was a 'head-hunting' call, and the first one I had ever had. They were planning to open an office in Singapore to spearhead an expansion into Asia and they wanted to know if I would be interested in heading it. It was flattering, of course, and I had one or two meetings with the senior management of CRA, but I would have needed a very large inducement to leave BP, and our talks came to naught.

Summer arrived and my weekend visits to Bundy Hill became even more frequent. Again I marked the New Year by a day at the Hanging Rock picnic races. And before 1974 was much older I began making my plans for returning to England for my home leave.

The landlord of my flat on the seafront at Middle Park had indicated he wished to return to live there when my two-year lease expired, and so I would have to find a new place to live. Just before departing on leave in mid-February, the packers came to take my belongings into storage, and I moved back briefly into the St Kilda Travelodge.

That leave was as full of interest and as memorable as any home leave I'd had. Every step of the way brought me new pleasures and so I shall describe it in some detail. I was now experienced in re-routing the air tickets issued by the Company, and I started by flying to Singapore. At the Swimming Club I lunched again with Dick Murphy and his wife, Kate; I met the Sipiere family and other old friends, and visited the BP office, discovering a whole new team there now, led by Peter Hardwick (whom I reminded with amusement of the brief helicopter adventure we once shared in Alaska). After a couple of days I flew to Bangkok and checked in to my favourite hotel, the Oriental. At the BP office, a stone's throw away, Geoff Larminie was away but my old friend Khun Dtao, who had taught me my first words of Thai, was there still. I wined and dined with other friends and spent some time at Pattaya, swimming and water skiing.

I was uncertain of the reception I'd get from Mary Lou in Manila, but she was as warm and forgiving as ever, collected me from the Hyatt Hotel when it was time to leave, and saw me off at the airport. Crossing the date-line I arrived at San Francisco and took a room at the Mark Hopkins. Struggling with jet-lag, I reached Casper, Wyoming, on the

28th February – in time to learn that Labour had just won the British General Election and that Harold Wilson was back in 10 Downing Street.

Paul Truitt and I had stayed in touch since the time we had both worked in Bangkok. He was still working with Gulf Oil and had written that he was now based at Casper where " ....we live up in the hills and most days deer come into the garden to feed." It was good to see Paul again and to meet Christine, whom he'd met and married while working in Argentina. They were both expert skiers and before long we were on the slopes together, although I was completely outshone – my own experience amounted to just a few weekends dodging gum trees on the nursery slopes in the Australian Great Dividing Range. But it was the Casper Mountain chairlift that I remember almost as fondly as the skiing itself, sharing rides to the top and getting to know the friendly local skiers who were so amused by my 'cute accent.' Before leaving the States I went shopping in Denver, so when I arrived at my parents' home at the beginning of March I was fully equipped with skis, boots and suitable clothing.

The ski season was now at its best in Europe and I put my new gear to good use, spending a wonderful ten days in the Arlberg. Based at the Hotel Krone in the Austrian resort of Lech, I joined the ski school and made good progress. By the end of the holiday I was tackling the black-run Valuga and the Albonagrat, and what I lacked in style I made up for with enthusiasm.

A Thai friend named Lucky was living in nearby Lichtenstein and she and her husband had invited me to stay with them after my skiing trip. I had attended Lucky's wedding in Bangkok while she was working temporarily for BP and, perhaps because I was able to give her moral support in the face of parental opposition to her marriage to Alfonse, a *farang*, we had become friends.

Looking back now, it is remarkable how little I knew of Europe in the 'seventies. My British contemporaries seemed familiar with every corner of the continent from their frequent summer trips or skiing holidays there, but for me it was virtually *terra incognita*. Apart from a hitch-hiking journey through France as a student, and attending a short course in oilwell-logging in Paris, I felt a stranger on the continent. And so I welcomed Lucky's offer to show me this German-speaking corner of Europe. Both at Lech and now driving around the region with Lucky, I was conscious of my near-total lack of ability in German. It was not a feeling I enjoyed, and on returning to Melbourne one of the first things I did was enrol with the Goethe Institute for an elementary course in the language. As for Lucky, in due course she left Alfonse and married a

Swiss-French man, Yvan, and settled in Geneva, adding a fourth to the list of languages she can speak fluently. We have remained good friends.

Back in Britain, while waiting for the girls' Easter holidays to begin I flew to Glasgow to spend a few days with Ian and Julia Rolfe at their new home at Dalry, formerly the home of the Ayrshire landscape painter George Houston. On returning, the first thing Helen and Sally wanted me to do was help choose a pony to replace Griffy who had died mysteriously. We settled on Tango, a dun pony of 14.2 hands, and while she was being re-shod and inspected by the vet we left on a short holiday together – the first time that Anne had agreed to my taking the children away without her. We stayed a few nights with John and his family at Worcester and then had a week of pony trekking based on a farm at Glyn Ceiriog in North Wales. The girls, of course, wanted the friskiest ponies available, but for me they managed to find a suitably docile animal, a horse named Major with a nasty habit of continually breaking wind. It was a very happy holiday.

A mildly upsetting tailpiece to this whirlwind home leave was a minor incident in Zurich, from where I'd arranged to fly back to Melbourne. Walking through the city with my bag in one hand and my new skis over my shoulder (for I expected to use them back in Australia), I was startled when a car caught the end of my ski package and sent me sprawling to the ground. Although I was on a pedestrian crossing at the time, the driver evidently thought I had no right to cause him to slow down, and had struck me intentionally. Inevitably, Zurich remains in my memory as a city of aggressive drivers.

Stopping in Singapore I was pleased that Mrs Melhuish was able to join me for lunch at the Hyatt Hotel. Although by now quite elderly, her ability to chat in Thai without seeming to draw breath was undiminished.

A priority on returning to Melbourne was to find somewhere to live. The place I settled on was a cottage-style townhouse, possibly of prewar age, in Kensington Road, a quiet and leafy street in South Yarra. I was delighted with it and took up residence in mid-May.

That weekend the Pannells had a party at their home in the Dandenongs and we stayed up much of the night as the results of another General Election came in, the second in only 17 months. Lacking a majority in the Senate, Gough Whitlam had called this snap election to break a deadlock over what the opposition considered was Whitlam's skulduggery in seeking to engineer a Senate majority. Again there was much rejoicing among the Pannells' friends, many of whom were academics, over another Labor victory, albeit still without a clear majority in the Senate.

The Pannells were unfailingly generous in their hospitality, and I remember with wry amusement one small dinner party they gave. Apart from the host and hostess and myself, the only other guests were a certain Howard Jacobson and his wife. In the course of dinner it emerged that Jacobson, like Richard, was an academic in the field of English Literature. It was a pleasant enough evening until we were well into the second bottle of wine. Then Jacobson, with increasing aggressiveness, launched into an attack on me – or specifically on the way I speak, my accent. I wasn't aware I had an accent, other than a fairly standard Southern English one, but Jacobson was from the North of England and accents were apparently important to him. To everyone else's embarrassment he became more vicious and unpleasant, accusing me of adopting an affected voice, of being a poseur. It was clear the dinner party was effectively ruined by his boorish behaviour. Tempted though I was to invite him to 'step outside' I decided eventually that it would be sensible simply to leave. While Jacobson continued in the same vein I apologised to Richard and Sujatha and prepared to depart. Then to my surprise, and I must say, pleasure, his wife got up too and announced she would also leave. I naturally offered her a lift home, which she accepted, and we disappeared together into the night.

That was the last I saw of Howard Jacobson. From time to time these days I read his column in the Guardian or the Independent, and recently my ears pricked up when I heard an interview with him on Radio Four's 'Today' programme. Talking about literary festivals he denounced them as a fatuous waste of time, and his tone was as belligerent as I remember it from thirty years ago.

Winter returned and I appreciated my snug cottage in Kensington Road. I wrote home on 31st July:

"I'm on the hearthrug in front of a log and briquette fire – briquettes are compressed brown coal, like those things we used to burn in the kitchen boiler, called 'ovoids.' It's a cold wet night and I've not long finished dinner. For several days I've been living off the remains of a 14 lb ham which formed the basis of a party I gave last week. It was to say farewell to Angus Findlay and Martin White, two geologists who are leaving. I had the whole of Exploration Division, and 35 people in this living room made for a strong feeling of togetherness.

That pesky car still isn't in my possession but at least it is now on Australian soil, after heading off to Sydney on a ship which called here but couldn't be unloaded. Really, if you think England is a hotbed of strikes you should see this place. Labour relations are awful, and hardly a day goes by that you are not affected by some strike or other. There is virtually no dock labour; today no caterers at the office (and oh!, how I missed my tea); all forms of transport will grind to a halt at the end of this week; and so on. What with these problems and a 16% inflation rate, the country is heading for real trouble. I don't know what to do for the best about inflation and what to do to hedge against it, but it certainly seems crazy to hang on to any cash. Ideally one ought to take on enormous debts!

How are the investigations into your dizziness going, Dad? By a remarkable coincidence I was at a farewell party a couple of months ago at Richard and Suja's place. It was for a young doctor called Michael something-or-other. I've since learned that he's based at Harold Wood Hospital, so maybe you'll come across him.

Last weekend I bought a few more plants and it is getting to the stage when I shall soon have to force a way through the vegetation in the front porch to get to the door. I got a couple more eucalyptus saplings, a muscatel grapevine (now there's something you ought to get for the greenhouse!), Antarctic beech, pink heath, waratah, and a couple of others I dug up in the countryside.

Yesterday, a day off (did I tell you about the oil industry's nine-day fortnight?), a couple of friends and I went skiing up at Mount Buller, 150 miles northeast of Melbourne – good fun but atrocious foggy weather. I'm beginning to doubt if Australian skiers ever see the sun!"

As spring returned and the autumn leaves in England began to fall, the flow of visitors from Head Office increased. If they were non-technical managers this simply meant meeting them for cocktails and canapés at receptions given by the top man in Melbourne, Michael Rendle. But if they were from Exploration Division I would generally find myself escorting them around our main areas of activity: the offices of Burmah Oil in Perth, and the PNG Highlands. I remember the visit of London's General Manager of Exploration, a dapper and talkative man named Roger Bexon. After presentations in Perth we were now on a flight back to Melbourne, and in view of his status we were in the front cabin, flying First Class. The stewardess was having difficulty opening a bottle of wine, and – ever the gentleman – I leant from my aisle seat and offered her a hand. I hadn't realised as I grasped the neck of the bottle that simultaneously she would press down on the arms of the corkscrew! The cogs on the demon gadget chewed into my flesh and I could barely stifle the instinct to scream. With sweat breaking out on my brow, I beseeched her through gritted teeth to lift the levers again and release my hand. She did, and I slumped back in my seat, wondering if my boss had observed the incident and, if so, how many points he'd award me for *sangfroid*.

Harry Warman, another top Exploration Division man, arrived in November. I liked Harry and we'd known each other from as long ago as my time in Alaska. I laid on a series of presentations of the work we were doing, but the matter I was most interested to talk about with him was his plans for my future. I'd been in Melbourne for thirty months and, while I would have been pleased to stay – notwithstanding the squeeze being put on foreign oil companies by the federal government – I knew that this was already longer than any of my previous postings and a move must therefore be on the cards. In a *tête à tête* over dinner at Frenchy's Restaurant downtown, he outlined his plan for me to take over as Exploration Manager of "the Consortium" in Iran. This would be a very large job and a huge increase in responsibility for me and so, although somewhat apprehensive, I was pleased at the prospect. A few days later I saw Harry onto his plane, and waited for further news of my Iran posting.

Meanwhile, I was increasingly concerned over the Labor Government's hostility toward foreign-owned oil companies. In a letter home I wrote:

> *"Ken Roberts is off to London next week to attend a top-management meeting on the northwest offshore operation. With the nationalistic government we have here, foreign companies such as the Burmah/BP/Shell/California Standard group off West Australia are being squeezed. The form it's taking is the government's unwillingness to discuss the price it (i.e. the monopoly buyer) will pay for the gas we are discovering. All the partners are getting thoroughly fed up with this and there is a growing feeling that exploration is hardly worthwhile since, whatever we may discover, we are not free to sell it as we please.*
>
> *So I may find when Ken comes back that my area of responsibility will have been reduced to Papua New Guinea! Up there, fortunately, we are pretty active and drilling will commence there in March. I've just finished putting the finishing touches to a proposal to drill a $4 million hole at Lavani, up at 8000 feet in the Highlands."*

With the summer, the cicadas were chirruping again at Bundy Hill and I would visit there whenever I could, to barbecue with a friend or two, to camp, or simply to potter there among the spring flowers, enjoying the rich bird life and the peace and quiet. But it was a party in Melbourne at the end of November which – though it may sound fanciful – was to change the course of my life. Rudi, who by now was viewed by

Melbourne's Thai community as their doyenne, decided she would hold a *Loy Kratong* party at her home at Sandringham. In Thailand *Loy Kratong* takes place on the night of the November full moon, when small and beautifully decorated rafts illuminated with lighted candles are floated down the Chao Phraya River as an act of purification, these tiny rafts – the *kratong* – bearing away all of one's sins and bad luck. The Thai women attending Rudi's party were asked to make a *kratong* each and bring it with them. It was a delightful occasion and as the evening darkened and the full moon appeared, the swimming pool in the garden became a picture of exquisite candle-lit *kratongs*.

I had asked Rudi if I might bring a friend, a girl called Sue, but as the evening progressed and I moved among the other guests I found I would keep returning to a certain young Thai woman, somewhat, I confess, to the neglect of Sue. She introduced herself as Song and explained that while that was her nickname, her full name was Thammajaree Dharmajiva, and she was a physiotherapist working at Caulfield Hospital. She had been in Melbourne a year or so and although initially she had stayed with her brother and his Australian wife, now she was in a flat of her own. Naturally, I did my best to impress her with my grasp of Thai and we talked of her work, her family and of the time I'd spent in Thailand. Before the evening was over I had Song's address and telephone number. For the record, it was almost a year to the day since my divorce from Anne had been decreed absolute.

Before I had a chance to meet Song again, a business trip took me away from Melbourne for several days. The 1975 conference of the Australian Petroleum Exploration Association was planned to take place at Surfers' Paradise and I was the Chairman of the Technical Papers Committee. It fell on me to plan the conference technical programme, to solicit papers for inclusion, to edit them and arrange for the volume to be printed so they could be distributed at the conference. While in Surfers' Paradise checking the suitability of the conference hotel, I found the time to write a postcard to Song, asking if she would have dinner with me the following week. She agreed, but in the course of the meal at a South Yarra restaurant she pointed out that, in her view, sending a girl a postcard of near-naked Queensland parking attendants, the so-called 'metermaids', was a strange way to invite a woman to dinner. Apart from that small hiccup our relationship progressed well.

Meanwhile, I had been corresponding with Anne to allow Helen and Sally to visit me. She insisted that they should spend Christmas in England with the rest of the Gallagher family, but shortly after that the girls flew to Melbourne for what was to be a very happy fortnight together. We visited the sights of Melbourne, picnicked at Hanging Rock for the New Year race meeting (for me now, something of a ritual), barbecued at Bundy Hill and camped in the Great Dividing Range. Shortly after I had seen them off at Tullamarine Airport I wrote to my parents:

> *"Well, the girls have come and gone and I hope by now they will have told you about their trip. They seemed to enjoy themselves mightily and I think they thought of it as a holiday. As for me, I preferred to think of it as a visit to their second home, and I tried to reinforce this idea in their mind by giving each a bedroom which they were to think of as their own. Anyway, the result was the important thing – we got on really well together and by the end of the fortnight I felt that the bond between us was every bit as strong as it's ever been. And we were even blessed by the weather which was warm and sunny right up until the day before they left. As you can imagine, after they had gone the house seemed pretty quiet and empty ...."*

It was now 1975 and my father went into hospital in January for an operation to remove the 'warts' from his bladder. I wrote to him and wished him a speedy recovery and took the opportunity in the same letter to tell them about the new woman in my life:

> *" ....It's a coincidence that you should have spoken of your physiotherapist, as I also have one. No, she doesn't give me ankle exercises or anything like that. She's a Thai girl called Song and has been working over here at the Caulfield Hospital. I met her at Rod and Rudi's place back in November. She's a typically soft and gentle Thai girl but, of course, well educated and from 'a good family'. She makes me happier than I've been with a girl for a long time (even though she won't let me sleep with her – Mum, you'd better not read that bit; men only, you know!) She's cooking me dinner tonight.*
>
> *Still no news of Persia. I wrote to Harry this week asking what is going on, as I'd received a letter from my old friend Ron Walters saying that over there its 'an open secret' that I'm being posted to Aberdeen. So naturally I'm a bit puzzled."*

The next letter from home came from my father and concerned matters more serious than where I was to be sent next. A biopsy following his recent operation showed that he was suffering from cancer. I sent off a reply immediately:

> *" .....well, it's a blow of course to find that you've got cancer, Dad, but I've got high hopes that the radiotherapy will fix it. Something*

that will surely increase your chances of it being successful is your lion's constitution. I gather that the first session is tomorrow – let me know how it goes. In the meantime, will you be at Harold Wood Hospital all the time or will you be going home?

Dad, your letter sounded pretty gloomy, but I suppose that you've got to face the possibility of the deep-ray treatment not curing you. While I think the chances of success must be darned good for a man of your physical strength and willpower, I also want you to know that if things should not go well, I shall of course come back and see you. Remember that such trips are paid for by the Company, so let there be no talk of 'I didn't think I was bad enough to trouble you.' Even if you're not returned one hundred percent to good health, rest assured we shall see each other again ..."

The rumours of my impending move to Aberdeen turned out to be correct. Apparently, Harry's lobbying on my behalf with BP's partners in Iran were unsuccessful and it was decided that the next Exploration Manager there should be a Shell appointee. 'Well, it's an ill wind ...' I concluded, and living in Aberdeen would at least allow me to see more of my parents at this difficult time, and of the girls.

In the weeks before I was due to leave Australia I made more visits to Perth for meetings with Burmah Oil, escorted visitors to Papua New Guinea to examine the preparations for the forthcoming drilling operations, and of course, continued to spend time with Song. The APEA Conference at Surfers' Paradise in mid-March was a success and I was complimented on the journal I'd produced.

And, importantly, I took Australian citizenship. Changes to the immigration regulations were afoot, but for the time being it was still possible for a British person to gain Australian citizenship with minimal residence qualifications and I took advantage of this, reaffirming my loyalty to the Queen. A few weeks later I obtained my blue passport to go along with my red British one.

On 3rd April I wrote to my parents:

"Tomorrow fortnight, Friday the 18th, will be my last day at the office and on that Sunday Song and I are off to Bali for a few days. After that we shall be flying to Bangkok to stay at her parents' house for a couple of weeks. You will have gathered from the tone of my last few letters that I am very fond of Song, and so it won't come as a surprise to learn that we may shortly be getting married. It's all uncertain still but she is keen that we should do the thing properly by my going through the motions of asking her parents first.

We are compatible in so many ways that I'm sure that with our maturity and good sense we could make a good marriage. Song is a quite incredible girl in many ways and one of the first things that attracted me to her was her strength of character and her ability to cope on her own in, what for her, was a totally alien environment. She is shy with strangers but I love her great sense of humour, her sense of responsibility and her naturalness. And, rather important, she seems to love me and wants to spend her life making me happy – now that I've persuaded her what a good idea that would be, and she wasn't easily persuaded.

I realise you won't be happy about the racial difference but, as you know, I've always believed in racial equality and hated any form of prejudice. Perhaps living abroad for so long and feeling to be 'a citizen of the world' (to use a cliché) has made me largely unconscious of race. Anyway, I hope we shall receive the blessing and goodwill of you both should we marry in the near future.

Well, that's my bombshell for the week. Mundane matters like what I'm going to do with the car can wait.

All my love ...."

David Balchin and his wife Gloria gave a splendid farewell party for us at their home, complete with a vast steaming haggis to prepare us for our new life in Scotland, and on Sunday 20th April we were waved off at the airport by assorted friends.

Nothing ground to a halt when I left! The Northwest Shelf project went on to become a major source of LNG for export to Japan. In Papua New Guinea our Lavani borehole and the boreholes that were drilled by newcomer Oceania were dry. As often happens after a string of dry holes, management decided to seek a partner to bear the cost of further work and so, in 1978, Australasia Petroleum Company farmed out its Southern Highlands licences to Niugini Gulf. With improved seismic technology Gulf was able to map the internal structure of the anticlines and found that the sandstone reservoir which we had been targeting was less complexly folded than I, at least, had expected. In a second well on the Mananda Anticline in 1983 Gulf found oil traces. A year later it found gas in the Juha Anticline, and in the years that followed there were further discoveries of oil and gas along the length of the Highlands. BP's erstwhile partner, the small company Oil Search, remained steadfast throughout and participated in many of these discoveries, gradually increasing the size of its share in the project until today it is the largest player in PNG and the operator of the venture – if ever there was a case of perseverance eventually paying off, this was it.

Chapter Thirteen

Song was not yet fully recovered from a brief stay in hospital and so our days in Bali, at the Tandjung Sari, and then our visit to Jogdjakarta and the Buddhist temple at Borobudur, were less energetic than they might have been. We flew to Bangkok a few days later and took a taxi to the family home in Soi Klang Akapat, off Sukhumvit Road. My future mother-in-law was sitting in the garden when the car drove up, and if I was expecting her to jump to her feet and run to greet her daughter, I was disappointed. To conceal one's emotions is, of course, a typically Thai feature, but to show so little interest, I thought, was shocking. But Song was unruffled.

Song's father, on the other hand, was someone I warmed to. Before retiring he had been a manager with the Bangkok Bank but now his life revolved around golf and his Buddhist religion. The Dharmajiva family had once been one of the wealthy families of Bangkok. Song's grandfather had been educated at Eton and Oxford University and had gone on to become the Ambassador of Thailand in the Netherlands. Their son (my future father-in-law) spent much of his childhood in that country before being brought up by an English family, although I never learned why he was farmed out in this way. The result was that he spoke English well, albeit diffidently. They had been big landowners and still had a large acreage near the town of Chachoenseow, east of Bangkok, as well as owning a printing works in the back alleys of the Chinese area of Bangkok's old city. But *Khun Por,* as I called my future father-in-law, was no businessman and in spite of his banking career he was slowly being cheated out of his land by an unscrupulous Thai partner with whom he had teamed up to develop a country club. Meanwhile, Song's mother seemed to be squandering what remained of their money on jewellery and *mah jong* gambling sessions with her friends. Song had many siblings but closest to her was Nai, a stewardess with Thai Airways, who in turn became a good friend of mine too. I settled into this ménage as best I could.

Song's father was insistent that the timing of our marriage should be based on the advice of fortune-tellers. Arrogantly I thought that this was ridiculous, and as the days passed and still no advice came from the *mor-doo,* I became cross and frustrated. But at last the date was decided. Apparently 30th April would be an auspicious day. Preparations for a marriage at the family home were put in hand. Song's mother – again to my annoyance – insisted on taking Song and me to her favourite jeweller, and intervened throughout as we chose the wedding ring. The wedding day arrived. Nine Buddhist priests in their saffron robes intoned prayers as we knelt with a sacred string connecting our heads, holy water from a conch shell was poured on us, and in a daze I did as I was bidden. With the ceremony over, the local Thai registrar recorded our marriage and issued us with a certificate. We were now man and wife, me aged 38 and my new bride 26. Food and drink followed, with the family seemingly content that the ceremony had been properly conducted. By happy chance, our friends from Melbourne, Rod and Rudi Hood – at whose home Song and I had met – were holidaying in Bangkok at the time and we were glad they were able to join our celebration. We spent our wedding night at the Oriental Hotel and on the following day read in the Bangkok Post that the Americans had finally withdrawn from Saigon on the day of our wedding. To have married on the day that peace descended on Vietnam was, I had to admit, indeed auspicious.

We did not have a conventional honeymoon. Song was keen that her old school friend, Lek, should accompany us for the few days we spent at Rock Cottages, on the coast north of Pattaya, and on the 13th May we boarded an Air France flight and left for Europe.

As ever, I was keen to take advantage of the opportunity to stop at new places en route, and we broke our journey first at Athens. The Adonis Hotel was in Plaka, the old part of the city, and on the first morning as the sun rose Song opened the curtains, stepped onto the balcony and, looking up at the ancient hill-top ruins, called to me: "Khun Mike, come and see this – a wonderful view of the Chiropodist!" Song spoke excellent English but was subject to an occasional malapropism; she saw the funny side of this one and laughed over it as much as I did.

They were digging up the streets of Vienna when we were there and we got to our hotel on Wenceslas Square by scrambling over banks of rubble. In the public gardens we fell foul of a local lady, a busy-body, who upbraided us for daring to walk on the grass. But we did attend a concert by the Warsaw Philharmonic in the fine baroque Musikverein, which Song said had long been an ambition of hers.

We stayed for a week or so with my parents at Harold Wood and my mother was charmed by Song. The two got along like a house on fire, going shopping together, talking endlessly, and my mother happily responding to Song's enquiries about my favourite dishes and how to cook them. I think my father was fond of Song too, although not endlessly patient with her as was my mother.

I was due to take up my new job at the beginning of June, and we took the 'Aberdonian' from Kings Cross, arriving in the 'granite city' on a grey and wet Saturday evening. We paid off the taxi outside the Skean Dhu Motel at Dyce and I'd swear there was sleet as well as rain in the cold northerly gale which buffeted us as we struggled with our luggage across the forecourt.

It was just a walk to BP's new and modern North Sea headquarters. Again, I found myself taking over as Chief Geologist from David Jenkins and, as I expected, everything was running efficiently. The whole UK offshore region had been divided into separate areas and geologists worked alongside geophysicists in these geographic teams, each team being responsible for producing proposals for exploration wells in their area, for planning the appraisal drilling on the oil and gas fields which had been discovered, and for proposing blocks of acreage the Company might apply for in future government licencing rounds. It didn't stop there of course, because in many of the operations BP was in partnership with other oil companies, so there was a constant round of meetings to agree courses of action which would meet the (often different) objectives and priorities of the different parties.

BP had several drilling rigs operating and as well as ensuring that none of them was allowed to stand idle – in other words that they had a new location to move to when the last borehole had been drilled – our job was to make sure there was always a competent geologist on board. This side of the business was called Operations Geology, and I was pleased to have my old friend in charge of this function, Stuart Buchan, with whom I'd worked in Alaska and in Singapore. Stuart's job was to train the new graduate geologists that BP was now recruiting in large numbers, to roster them on the drilling rigs, and to see that they had the necessary technological backup in their offshore laboratories so that the quality of their work was of the highest standard. It was a demanding job which involved working unsocial hours and Stuart, still a bachelor at that time, was ideal for it.

Since BP had discovered the first gas field in the Southern North Sea, the West Sole Field in 1965, there had been a string of further gas discoveries in that basin. In the Central and Northern North Sea most of the billion-barrel-plus 'supergiant' oilfields were discovered in the early 'seventies, before my arrival – fields such as Forties, Brent, Ninian, Claymore and Magnus. One-hundred-million-barrel-plus 'giant' fields were still being discovered in the mid-seventies, but the prospects were becoming more subtle and the fields therefore harder to find. Meanwhile those earlier oil discoveries were being developed – that is to say, brought to production – and at Dyce we had a large team of engineers overseeing this technological challenge. Shortly after my arrival at Dyce the Queen paid us a visit to inaugurate the first of the supergiants, the Forties Field, and we all came away with a souvenir bronze medal commemorating the occasion. That early phase of very large discoveries followed by diminishing sizes is typical of successful exploration in a new geological province, and I remember my boss, Peter Walmsley, saying: "We've picked up the fivers and the tenners, and now we're busy picking up the loose change!"

The Dyce office was the biggest and most important of all of BP's operations worldwide, with the possible exception of the Alaskan venture. The team of geologists numbered a couple of dozen, among them several I had worked with elsewhere. But there were in addition a lot who were new to me. One was Chris Gibson-Smith, a protégé of David Jenkins, whose geological output was unexceptional but who possessed a certain charm and smoothness of manner that most of us geologists lack. It was therefore no surprise to find, long after I'd left BP, that Chris had progressed up through the ranks of management to the main board. From BP he moved on and, as I write, he is Chairman of the London Stock Exchange. While I'm basking in the reflected glory of others, I might mention that John Browne also spent some time at Dyce while I was there. Occasionally we would attend a meeting together or share a taxi. He was a petroleum engineer, but already showing himself to have more commercial acumen than most of us, and he continued his fast-track career to become Chief Executive of BP, for which he was elevated to the peerage as Lord Browne of Madingley.

At the bottom end of the experience spectrum, new graduate geologists were being recruited in larger numbers than BP had ever recruited before, and most of them were sent to Dyce to begin their careers. As yet, they were all male since it was considered unacceptable to allow young women to work in the rough and tumble environment of an offshore drilling rig. But that was soon to change. These young geologists matured as oilmen much more rapidly than I or my contemporaries had done, partly because of the excellent training that they were now getting within the Company, and partly because at Dyce they were able to experience the full cycle of exploration in the course of just a couple of years – from the initial application to the government for a block of exploration acreage, to the seismic surveys, to the first exploration borehole and, if successful, to the appraisal drilling to confirm the discovery's commercial viability. Little wonder that from their ranks came the chief geologists, exploration managers and managing directors of a number of oil companies around the world.

Away from the office, Song's and my priority was to find somewhere to live. There were plenty of properties to look at and we soon realised that the more agreeable areas were to be found in the valleys of the Rivers Dee and Don, where there was some protection from the often harsh winds which blew in off the North Sea. Song was emphatic that we should make an offer on a detached house on North Deeside Road, by the name of Eastwood. I went along with her preference reluctantly. Inside it seemed to lack space, the external paintwork was an unpleasant pale blue, and the overgrown garden blocked any view of the Dee Valley. Our offer of £25,500 was accepted and we took possession in early August. Song's proposal was that we should drastically change the internal layout, by removing certain dividing walls and enlarging the kitchen. Although I argued that it would be too costly, we went ahead and several months later Eastwood emerged from the rubble, a spacious and attractive house with a large and modern kitchen. By then I had repainted the exterior black and white, and carried out major surgery on the garden to open unrivalled views southward across the Dee. Song had proved that where vision and imagination were concerned, she had them in abundance.

Her employment experience was far less happy. Song had obtained a degree in physiotherapy at Bangkok's prestigious Mahidon University, and when she went to Australia her qualification and working experience had been completely accepted, allowing her to work alongside, and with the same status as, her Australia-trained colleagues. In Aberdeen she was dismayed to find that things were different. She was employed in the city's main hospital only as a physiotherapy assistant, and moreover, her boss, Miss Rosemary Lane, was a bullying harridan. Song would return home of an evening, grumbling about "that ghastly woman," and about the old-fashioned British approach to physiotherapy and the poor status of physiotherapists, even of those colleagues who were fully qualified.

Aberdeen is not a city in which I would have chosen to make a home. While it was clean and prosperous and the oil industry was encouraging amenities such as hotels and restaurants to multiply, we found the native people to be inward-looking and we made few friends among them. But we did enjoy certain things it had to offer. The countryside up the Dee Valley is attractive and we soon joined the National Trust for Scotland, becoming frequent visitors to the many castles and other ancient monuments of Northeast Scotland. The summer of 1975 eventually brought warm weather, and by good luck our house was far enough inland to be free of the North Sea mist, the *haar,* which on fine days we could see as a grey wall in the east. Our garden was of a good size and Song and I enjoyed working in it, building stone walls, getting old hedges under control and developing a useful vegetable patch. And in the winters we would drive through Ballater and past Balmoral to Glenshee where we braved the Arctic conditions and Song soon picked up the rudiments of skiing.

But those first years of our marriage were difficult, and too often we were at loggerheads. Song's strength of character was admirable, but she had an uncompromising streak which I found difficult to accept. She had a way of using the Thai expression *mai chorp,* meaning 'I don't like it,' whenever I expressed a preference which was not in line with her own. Admittedly, often her preferences did turn out to be more sound than mine, for example, over the choice of a house, but it was aggravating always to have my own views discounted. That friction between us was as hard on Song as it was on me, but for her there was also the thought that she had given up a pleasant and fulfilling life in Melbourne for a frustrating job in Aberdeen.

We had money problems too. Gone were the expatriate's perks I'd enjoyed over the past years, as Aberdeen was quite reasonably deemed a home posting. And a government-imposed pay freeze and inflation running at over 25 percent only added to my difficulties. To make matters worse, my ex-wife was making financial demands I found hard to meet, and at one point I was threatened that bailiffs would descend on our house.

There were times when Song would become withdrawn and uncommunicative and I'd be desperate to restore her to the person I'd known in Australia. Just what prompted me to write what follows, I cannot remember; perhaps by describing in detail this particular incident I thought it would help me to understand our predicament. I wrote it in pencil on a few sheets from a note pad, and although it is undated it was in effect an extended diary entry:

"Song is sitting in the bedroom, cross-legged, with her head in her hands, sobbing. We sat and watched 'The Money Programme.' I

asked if she wanted a cup of tea, and when I came back from the kitchen she was sitting on the hearth, crying. Try as I may, she won't utter a word, so I don't know what the trouble is.

She has been, in a way, tense all day. In the usual way I made my breakfast and sat eating it alone. At ten minutes to eight, when I'd shaved and dressed, she appeared in the kitchen in her coat and said: 'I'm ready, it's time to leave.' Usually we exchange the odd joke or comment on the weather but she began the day in this abrupt way.

Because she wanted to see Miss Lane about her work and her future at the City Hospital, and a possible week off in March, she had the car today. At lunchtime she phoned. Pleasant enough she seemed, and asked if she could have the car on Monday, as Miss Lane is out of town till then. 'Of course,' I answered, probably in Thai, and asked if she had bought the tickets for the ballet tonight, or if she'd prefer to go out to dinner – Indian food. She'd think about it, she said, and we chatted a bit more and I reminded her to pick me up at the office.

This evening as she sat in the car, I saw she was glum. And glum she has been for the remainder of the evening. She cooked fried rice, but wouldn't chat as we ate it, just answering me when she had to. Then, when I volunteered to do the washing up, hoping it would improve the situation, she insisted on doing it herself..

So there we are: another bout of depression, homesickness, schizoid behaviour. Whatever it is, it is not something new.

Earlier in the week we had a flat patch, but as the days passed it seemed to have been forgotten. Or, at least, its effect had seemed to lessen. That had been over Song's unhappiness here in Aberdeen. She had said something like: 'The only thing that gives me any comfort is the thought of leaving this country.' It was an old familiar tune but it piqued me more than usual, and in exasperation I replied: 'For Christ's sake, can't you just accept the place? How in god's name are you going to adapt to our next posting if you can't adapt here! Egypt, Teheran or even Stavanger would be far more difficult to adapt to than Aberdeen!'

She retreated into quietness. I tried to recover the situation with normal trivial chatter as we continued our car ride to work, but she would only say: 'If that's really what you think, then there's no hope for us.'

That's the kind of incident that habitually mars our life together, because she takes them so to heart. After each time it takes a good two days of approaches and banter before I have her smiling again.

I heard her go into the bathroom and shower, (showering for Song has always been a kind of mental and physical therapy – a real Thai). Later I went back to her. She was no longer crying and I sat on the bed beside her, put my arm around her shoulders, and asked her what was behind her present breakdown. She forced herself to speak. 'What can we do?' she whispered, looking down at her hands. It was clear that no particular incident had prompted her crying, but it was the realisation again that we were unsuited to each other. 'We are so different' she said, 'we think differently in so many ways.'

We soldiered on, sometimes happy together while at others the black dog of depression would sink its teeth into Song's contentment.

She would take any available opportunity to escape Aberdeen. Her cousin, Dtun, was married to the Commercial Secretary at the Thai Embassy in London and Song liked to stay with them in their flat on Putney Hill. Then in early 1976 she decided to return to Bangkok for a while. We wrote to each other often while she was there and gradually the tone of her letters became more affectionate, and I sensed she was returning to her former self.

I felt I also needed a break and I headed for Verbier, spending a week at the Hotel Eden and enjoying warm spring weather and ideal skiing conditions. Feeling completely renewed, on the flight back to London from Geneva I fell into conversation with the young woman seated next to me. Her name was Lourdes, she was Bolivian, and she was completing her studies at an international college in Lausanne. She was not just attractive, she was enchanting. We seemed like soul mates and stayed in touch for several years after that first meeting. And then one day she told me through tears that her father, back in La Paz, had just been murdered. He was, I knew, an educated and wealthy man, but whether he was killed in a random attack or for political or drug-related reasons Lourdes chose not to say. She had now to return to South America, but she continued to write, inviting me to come one day and see her country for myself, " .... *porque Bolivia es muy diferente!*" Just how different it is, I have yet to discover.

After two months away, Song arrived back in Aberdeen. She had timed her return for our first wedding anniversary. The tulips we had planted at Eastwood the previous year were opening, and just a week later the first swallows returned. While she had been away she had decided that she would no longer suffer the indignity of being a physiotherapy assistant; she would sit the examination of the Chartered Society of Physiotherapists. She continued working by day, albeit as a physiotherapist assistant, and of an evening she studied at home. Growing piles of medical textbooks appeared around the house, with such titles as: *"A Handbook of Fractures," "Physiotherapy in Orthopaedics"* and *"Principles of Neuromusculoskeletal Treatment and*

Management". Toward the end of the year she sat and passed the examination and so became a Chartered Physiotherapist.

My work took me to London frequently, for meetings with other companies or with the government's Department of Energy. There were visits to Norway (which Song enjoyed), and to Teheran for work-programme discussions with the state oil company, NIOC, who were our partner in several North Sea blocks. And then there were various oil industry committees I was expected to join, organizing conferences or discussing matters of common exploration interest. That summer Dyce had a stream of visitors wishing to see something of the North Sea operations, and I remember trips to the Forties Field with dignitaries including Members of Parliament Russel Fairgrieve, John Biffen, Christopher Tughendat and, the most agreeable of the bunch, Hamish Gray.

I mentioned earlier that the geologists worked in teams with geophysicists. The latter investigate the physical attributes of the crust to determine the nature of its rocks, and while some geophysicists enter their profession having had a geological training, there are others who were trained in physics and so have less understanding of geological principles. I suggested to the Chief Geophysicist, my counterpart, that I should run a geological field excursion for those of his staff who were interested, and he enthusiastically agreed. I chose to take them to the Pennines and adjacent parts of Northern England, where over twenty years before I had been on my own first geological excursion as a schoolboy. For four days we travelled the area by hired coach, examining the succession of rocks which are so well displayed there, from the granite 'basement' at Shap, through the slates and volcanic rocks of the Lake District, the Carboniferous beds of the Craven district, and finishing the trip on the Yorkshire coast where we examined the Jurassic and Cretaceous beds which are so important in the North Sea oil province. I had done little or no teaching before but I found I enjoyed it.

In 1975 the Labour Government, wishing to ensure that the British nation should enjoy a larger share of the benefits of North Sea oil production, set up a new body which it called the British National Oil Corporation, or BNOC. Every oil company operating in the North Sea was obliged under the new legislation to enter into negotiations with BNOC to enable the latter to 'participate' in its oilfields, and I need hardly say that within the industry BNOC was not greatly liked or admired. One of its aims was to have an exploration arm of its own, as well as this so-called 'production participation' responsibility. And so to get the new company off the ground the government had rolled the exploration division of the National Coal Board (already involved in a number of Southern North Sea gasfields) together with the UK exploration and production side of the Burmah Oil Company (recently acquired by the government after being brought to its knees financially by over-extending itself on its downstream activities). The Chairman of the new company was Lord Frank Kearton, a cross-bench peer who had previously run the textile giant, Courtauld. To advise him on exploration, Kearton brought in my friend and erstwhile colleague, Harry Warman, who by then had recently retired from BP.

Early in 1977 I received a call in my office from Harry. He was aware of a certain key well that BP had recently drilled and because it was relevant to a particular operation of BNOC's, Harry was fishing for information. The geological information from wells is routinely kept strictly confidential because of the multi-million pound cost of acquiring it and therefore its value for exchanging with other companies – what are called 'well trades'. Nevertheless, a certain amount of hinting at a well's findings does take place between competitors on a *quid pro quo* basis. On this occasion Harry and I circled around each other, each giving away as little as possible while trying to find out as much as possible about the other's boreholes. When we were finishing our conversation I asked Harry out of idle curiosity if BNOC had yet filled the 'Chief Geologist' vacancy it had been advertising.

"No, not yet;" he answered, "why, are you interested?"

Until that moment it had not crossed my mind. But now that I was being asked, I replied:

"Yes, I daresay I could be."

As we continued to talk, the thought grew in my mind that perhaps it would make sense for me at least to consider it.

The fact is I was feeling unsettled, both domestically and at work. I had asked the new Chief Geologist in London to be posted overseas again as a solution to my financial difficulties, but had been turned down. The new job would pay more than I was then getting and would provide a way around the Government's pay freeze. Philosophically I was sympathetic to the aims of a state oil company, working in the interests of the nation rather than its shareholders. Being in on the creation of what could become a major oil company had its obvious attraction. And I'd been with BP for twenty years – a nice round number – and so perhaps it was time for a change.

A few days later I received another call from Harry to say that Kearton would like to have a chat with me. I talked it over with Song that evening and although she could see little merit in swapping my present job for what seemed to her a similar one based in Glasgow, I went to London.

As soon as I entered Kearton's office in Stornoway House, I was impressed. Although in his mid-sixties he seemed to have the energy and alertness of a man half his age. And having started his career as a chemist with ICI, I warmed to him as a fellow scientist. We sat and talked for perhaps an hour, and by then I was infected by his enthusiasm for the job in hand and by his obvious commitment to do what he strongly felt was in the interest of the country. By the time I arrived home I had decided that I should pursue this new opportunity.

My next meeting was with Alastair Morton. When Kearton had been Chairman of the Industrial Reorganization Corporation, the Labour Government's vehicle for its interventionist approach to ailing industries, Morton had been his protégé. It was clear where their political sympathies lay, and now Kearton brought him into BNOC to raise the massive bank loans needed to develop its share of North Sea oilfields. But Moreton's role was much wider than that, and so there I was in his office at seven o'clock in the evening, negotiating terms and conditions of employment with him. Eighteen months younger than me, bearded, and over six feet tall, he was an imposing figure and exuded dynamism and confidence. And I don't mean self-confidence in an arrogant way (although as I got to know him better I found he could be arrogant too) but confidence that however big the challenge it could be overcome.

Over the following days we agreed what I considered an attractive package of pay and benefits, and that I should start work forthwith. I explained that I was required to give BP three months' notice and, typical of Morton, he said that once I had submitted my letter of resignation he would contact BP and ask them to waive the notice period. Of course, BP was not about to do any favours for BNOC and so I had to take 'gardening leave' until my full notice period was up.

My colleagues were astonished at what I'd done, and couldn't understand why someone whose career appeared to be progressing well should jump ship. And I admit that at times I also wondered if I'd done a sensible thing. But there could be no going back. The die was cast.

My swansong with BP in those last few months was to put in hand a review of the oil prospects of an area lying west of the Shetlands known as the Rona Ridge, where BP and its partners held several blocks of acreage. Seismic surveys showed this geological feature to be an up-faulted ridge of old rocks buried beneath a thick blanket of younger sedimentary rocks, resembling an anticline – and, as geologists know, oil tends to migrate through sedimentary rocks and accumulate in anticlines. But further seismic surveys and the results of some scattered boreholes suggested to the men I'd ask to carry out this new review that the Jurassic rocks which were hoped to be present were probably absent – the ridge was probably composed of much older rocks over which

Cretaceous claystone was draped. The report was pessimistic and recommended that no drilling on this major ridge could be justified. My job as Chief Geologist was to write a critique of their work to accompany this new report to Head Office and to our partners. While I agreed that Jurassic rocks would probably not be present over the buried ridge, it seemed to me that provided they were present down the flanks of the ridge, and provided they had generated oil (in the way that Jurassic claystones are known to have generated much of the North Sea's oil), that oil would probably have migrated into the more ancient rocks forming the ridge. I disagreed with the report's findings and recommended that a well to test this hypothesis was justified.

The well was drilled later in 1977 but by then I had left BP. It found a large oil accumulation which was named the Claire discovery. Although the nature of the fractured Devonian and Carboniferous reservoir rocks and, paradoxically, the shallowness of the oil accumulation, meant that developing the field had to await advances in offshore technology and higher oil prices, Claire was eventually brought on stream in 2005. In terms of oil-in-place it is still the largest accumulation to have been found in the entire UK offshore area – although because of the complexity of the geology only a fraction of that will be recovered. Exploration successes are generally the result of a team effort, and Claire is no exception. I do, however, get some satisfaction from knowing that without my intervention it might have been years later that the well was drilled and the discovery made.

In the early summer, while waiting to take up my new position in Glasgow, Song became caught up in an unpleasant legal issue. It started one sunny weekend when we were taking a leisurely walk along the footpath which follows the line of the dismantled railway which used to take Queen Victoria up the Dee valley to Balmoral. The footpath ran through fields and woods between our house and the river, and was a favourite place of ours to wander. Since Dr Beeching's malign axe fell in the 1960s and the rails were removed, the track had become overgrown, and as we strolled we would stop now and then to pick wild raspberries or to focus binoculars on an interesting songbird or a hovering kestrel.

Two days later I was reading the Aberdeen Press and Journal when a report caught my eye. Apparently, on the same day as Song and I had taken our country stroll the body of a young woman had been found dead and sexually assaulted on the very footpath we had walked. Shaken by the news, Song and I cast our minds back. I had only the faintest recollection of seeing someone, a young man, but Song remembered in detail that same person. She said he'd squeezed past us on a narrow section of the path while I was bird-watching, and moreover she could remember in some detail what he looked like and what he was wearing. I immediately called the police. "What you're saying is most important" the officer said, "we shall send someone round to speak with you straight away."

Within a half-hour a police car drove up and a pair of detectives came into the house to listen to Song's account of our morning ramble. Apparently the murdered girl was an inmate of the nearby institution. As is normal in similar situations it seems, the police had immediately swooped on a suspect: a young man recently released from prison and with a history of assaults on women. Song's description fitted the man perfectly and it was established, therefore, that he had been on the track at around the time of the murder. A few days later she was asked to attend an identification parade at Aberdeen Jail – itself an unsettling experience for someone who, as an Asian, is visually distinctive – and she had no difficulty in pointing out the person she'd seen on the footpath. Song was now the key witness in the case.

As the authorities made their preparations for the alleged murderer to stand trial, Song was asked to attend several meetings at the Aberdeen office of the Procurator Fiscal. At the last of these meetings the official, a young Scottish lawyer, ran through the procedure that was likely to follow in the weeks ahead. She then asked Song to confirm that she would still be available as a witness when the trial was expected to take place. In reply, Song said, perhaps too light-heartedly: "Well, actually, I had been thinking of going to Bangkok some time this summer." It was the worst thing she could have said. Suddenly the atmosphere of the meeting changed. Instead of asking Song how firm were her plans, and whether such a possible trip could be deferred, the Procurator Fiscal became officious and shortly afterwards the meeting closed.

The next day a police car arrived at our gate for the second time. A Detective Inspector and a constable emerged and strode to the front door with a warrant ordering Song to appear at the Sheriff's office the following Monday. Astonished at this turn of events, we appeared in front of the Sheriff and the Warrant was read out to us, commanding my wife " ....to say why she should not be committed to prison" to ensure that she could not leave the country until the authorities had finished with her. To say that I was outraged would be an understatement. Only after Song had handed over her passport and I had made a deposit of £20 (in effect, standing bail for her) did she avoid having to go to prison until the trial had taken place. Back home that evening I busied myself writing to my Member of Parliament and to the head of the Procurator Fiscal's office to denounce this disgraceful treatment of an innocent member of the public who had become the victim of official stupidity for no other reason than she had behaved as a responsible citizen. Ironically, in the same week as this public-relations disaster was being played out, the Master of the Rolls, Lord Denning, was calling for greater help from the public in the Police's fight against crime.

As things turned out, when the alleged murderer came before the judge he pleaded guilty and was sentenced to life imprisonment. Song was therefore spared the final ordeal of telling her story in court.

Chapter Fourteen

When BNOC was established in 1975 the personnel comprising its skeleton management team were housed in London. A presence there was maintained throughout the life of the Corporation, but at an early stage the Government decided that the main office should be in Scotland, no doubt to appease the growing feeling North of the Border that the new-found oil wealth was Scotland's. It was said that Glasgow was chosen – as opposed to what would seem a more obvious choice, Aberdeen – because more than most places it would benefit from a large injection of well paid professionals.

Leaving Song in Aberdeen, I drove down to Glasgow on the first Sunday in June, checked into the grim Central Hotel, and reported the next morning at BNOC's offices at 150 Saint Vincent Street. Among the neglected but once grand Victorian sandstone blocks this, the Scottish Amicable building, stood out for its clean modern architecture. My boss was Dick Fowle, the Exploration General Manager. Along with so many others with whom I was to work, Dick had been with Burmah Oil. He was a courteous and friendly person, neatly turned out and, like myself at that time, he sported a moustache – in his case, a genuine hangover from his days in the RAF. He made me welcome and introduced me to the others in the Exploration management team.

It was an unusual arrangement. Reporting to Dick were three managers, all of equal rank: they were the Exploration Manager, Tony Challinor, not much more than five feet tall with a mop of frizzy hair and a jutting chin, but a capable man; Keith Tyrrell the Chief Geophysicist, a pleasant but rather ineffectual man; and then there was me as Chief Geologist. As far as my work went in the first year or so, it was similar to the work I'd been doing in Aberdeen, but the reporting arrangements meant that there was a permanent tension between the Exploration Manager and myself. There were also clashes of culture because of the preponderance of ex-Burmah Oil people with whom I was working. Some of these alien cultural differences I accepted, whereas others I managed over time to change to the BP approach.

I soon moved out of the hotel and the Corporation put Song and me in the Royal Scottish Automobile Club, a slightly shabby but centrally located establishment much frequented by Glasgow businessmen at lunchtime. In the evenings we were often the only persons in the dining room, unless Alastair Moreton was in town – he didn't relocate to Glasgow like the rest of us, although he seemed to spend at least part of every week up there.

During my enforced idleness in the early summer of 1977, Song and I had carried out a reconnaissance of the Glasgow region to assess the housing market and to get a feel for the relative merits of different areas. Driving south from Aberdeen we had approached Glasgow via Stirling and we immediately liked the countryside lying north of the city. The sale of our house in Aberdeen had gone well and with money in the bank we set about looking for a house to buy within commuting range of Glasgow. Over the next four or five months we must have looked at two dozen properties and had surveys carried out on many of them before submitting our sealed bids, as is the Scottish way. In some cases we were simply outbid, in others the vendor changed his mind about selling, and in one case the vendor died on the day we submitted our bid. It was all very disheartening. Meanwhile we continued to live out of suitcases at the RSAC in Glasgow, and BNOC's Personnel Department began to wonder if they would have to continue funding our 'relocation package' in perpetuity. Song became increasingly frustrated and despondent, and after a lot of discussion she decided that she would let me to get on with the house-hunting on my own. She would give it another month and then, if we still had not bought a place, she would return to Bangkok.

It had become a weekly routine to scan the Glasgow Herald property pages, and one day in late September, shortly before Song was due to leave, we came across an advertisement which caught our attention. It was for a derelict cottage with about ten acres of land in west Stirlingshire, near the southeast corner of Loch Lomond. It was called Millfaid. We obtained further details from the estate agent and that weekend Song and I drove out to look at it. After gingerly crossing a dangerously decayed footbridge over the Catter Burn, we picked our way across a meadow and up a small hill to the simple stone cottage. Most of its roof had been stripped of its slates, the doors were missing, and standing inside were half a dozen cattle, up to their knees in muck. It needed a major leap of imagination, but we decided we would try and acquire it. True to form, our initial offer was rejected. We offered the same sum but for a smaller area of land, and again we were turned down. The weeks went by but our efforts to negotiate a purchase were fruitless. Despondency again settled over us, and at the beginning of November, Song flew off to Bangkok.

A week or two later, with the prospect of buying Millfaid now looking remote, I was in my solicitors' office discussing another property. By now I was a familiar sight in the corridors of Maclay, Murray and Spens and, as my solicitor and I were talking the senior partner, John Spens, put his head around the door and asked how my house-hunting was progressing. We explained my lack of progress and as he listened, John's interest increased. He lived nearby, in the village of Gartocharn, and said that as far as he knew Millfaid was owned by Sir Hugh Fraser, the wealthy owner of Harrods, who lived a short distance along the lane from Millfaid. He undertook to speak with Sir Hugh's Glasgow solicitors. A few days later we met again. John had confirmed that Sir Hugh was indeed the owner of Millfaid – he'd bought it intending to build a house there for his helicopter pilot – but the party selling it was not Sir Hugh but a businessman from south of the border who had an option to buy it. The option was due to expire later in November when, according to John, Sir Hugh would accept a sum of £20,000 if we would care to submit a forward-dated offer to his solicitor. This was excellent news, and I instructed my solicitor to submit the offer.

As the option expiry-date approached I was confident that at last things could not go wrong, and the long house-hunting saga would soon be over. And then, with just a few days to go, I caught the name of Sir Hugh Fraser in a news item on BBC radio. What horror! He was involved in a major financial crisis. I no longer remember the details, but it was obvious to me that Sir Hugh would be preoccupied sorting out his problems and would not give a second thought to my offer sitting on his solicitor's desk. I spent a miserable few days expecting to have to resume house-hunting. But to my surprise and enormous relief the offer was accepted and, under Scots law the deal was binding. Perhaps Sir Hugh's problems were so great, I thought, that even my £20,000 offer was not to be sneezed at. Delightedly I sent a telegram to Song who, by now, had travelled on to Melbourne where she was staying with Rod and Rudi: "*Congratulations!*" it said, "*You are now a farmer's wife. Our offer for Millfaid has been accepted. Love, Mike*"

Song was in no hurry to return as by now she had resumed work in the Physiotherapy Department of Melbourne's Caulfield Hospital. I had in my mind an outline plan for developing Millfaid but it was going to be a large project and would take several years. We needed somewhere to live while work was underway, and I bought a flat in a modern development called Netherblane in the village of Blanefield. It was late January 1978 by the time I got possession and moved our furniture in from storage. Meanwhile I continued living at the RSAC.

Song and I corresponded frequently and I telephoned her to ask how she was faring and when I could expect her back. But the memories of her unhappiness in Scotland were still fresh, she was happy in Melbourne, and she continued to put off talk of returning. Christmas 1977 came and I accepted an invitation from my good friends the Rolfes to spend the holiday with them in Ayrshire. It was a jolly time and Ian and Julia were, as ever, generous hosts. A year or so later Ian told me that his widowed mother, Olwen, became a devoted fan of mine that Christmas. With dinner over, coffee was being prepared and someone expressed a wish for cream in their coffee. I suppose I'd had too much champagne, but that phrase prompted me to go down on one knee in front of the old lady, gaze up into her face, and croon the first few lines of the old song "You're the cream in my coffee ...." It was a ridiculous, impromptu, prank but apparently no one had ever done anything remotely like that to her before!

I often had to make business trips to London and would generally manage to squeeze in a visit to my parents at Harold Wood. I saw as much as possible of the girls too, and since Sally was now 16 and in the senior school, and Helen was at Freubel College in Roehampton, Anne was less able than she had been to put barriers in our way. During our sojourn at Aberdeen the girls had made a couple of short visits, but there was an underlying jealousy between Song and the girls. I decided now that I would invite them to come with me on a skiing holiday. Sally had made other plans, but Helen came with her friend Melanie and I booked a couple of hotel rooms at the Italian resort of Courmayeur. Our luggage was lost on the outward journey, there was a power cut on the first night at our hotel, and the weather was mostly poor. But the girls had skiing lessons and made good progress and, as for me, the poor snow conditions were good compared with Glenshee, and I enjoyed myself. Then back at our hotel on 24th March, relaxing after a hard day on the slopes, I received a call from my father. His voice was choking with emotion and I could tell immediately that something very bad had happened. He told me my mother had died that evening of a sudden heart attack. After a pleasant day out in the car, enjoying the spring sunshine, they had returned home and were sitting watching the evening news on television when she died. I returned with the girls as soon as I could find seats on a plane.

My mother was buried beside my sister Judy in the family grave in the shadow of Hornchurch church. She was a fine woman and the kindest of mothers, and how I regret that I had so rarely put my arms around her and told her how special she was.

I am not at all superstitious but I must comment on the timing of my mother's death. Her favourite brother had been Bob, a handsome and jovial man who on leaving Shrewsbury School had trained to become a civil engineer. After the Second World War, in which he had served as an army Captain and went through the evacuation of Dunkirk, he joined the civil engineering firm, Wimpey, and spent many years in Kuwait and Brunei. He was a hero of mine as a child, as well as my mother's, and I still have some of the treasures he brought back for me from the East, including a Dayak blow-pipe he had acquired up-country in Borneo. As I grew up and began my own career, my mother used to say how alike were Bob and I, not just in our similar careers and interests but even in our personalities. Sadly, Bob died of cancer when only 42 years old. My mother grieved sorely for the loss of her favourite brother, but additionally his death caused her a further worry. Because in her mind I was Bob's double in so many respects, she harboured a fear that for me also, 42 would be a dangerous age. In the event it was not I who died in my forty-second year but my mother.

My mother and Song had grown fond of each other in the two years since we had been married, and Song was terribly saddened by her death and by her inability to be present for her funeral. It seemed to bring about a change in Song, and she returned from Australia just a fortnight later. She approved of the flat I'd bought in her absence, and at last she seemed to accept the life that was taking shape for us. Glasgow was more to her liking than Aberdeen, the people were friendlier, there was more in the city to interest her, and she made a number of good friends among our neighbours at Netherblane and in the wider community.

My father had been dependant on my mother in all domestic matters. But being the man he was, he rose to the challenge of fending for himself. I would often call on him and find he had just finished the ironing or the cleaning, or else I would find he was preparing a full traditional meal for himself, with roast meat, vegetables and gravy, and a pudding to follow.

Since buying Millfaid, my plans for building a vehicle bridge across the burn to give access to the site had progressed, and I invited my father to come to Scotland and help me install it. John came too and so we had a 'full house' at the flat in Netherblane. Because the Catter Burn was the county boundary the bridge was to have one foot in Dunbartonshire and one in Stirlingshire, and this meant planning consent had to be obtained from both. Ian Rolfe had helped me survey the bridge site, and the bridge itself had been designed by an engineer colleague of my uncle Eric at the Forestry Commission in Edinburgh. On each bank a deep concrete foundation would have to be made and onto these would be bolted eleven-meter-long steel 'I' beams connected by cross-beams. Finally it was to be decked with railway sleepers.

In mid-May a local contractor arrived with his JCB excavator. Crossing the burn by the original ford, he dug out a deep trench on the Stirlingshire side and then returned to excavate the second trench beside the public road on the Dunbartonshire bank. I had worried about how we would get several tons of concrete mix across the burn to form the foundation on the Stirlingshire side, but a concrete pump was the answer and it was a satisfying moment when the pump's long elephant-trunk reared across the alder trees lining the burn and started dumping slurry into the first of the excavations. Small removable wooden boxes formed depressions in the top of each foundation and into these the holding-down bolts would be grouted once the beams were in position. All in all it was a fairly demanding engineering project and I was glad to have the advice and active help of John and my father to ensure it went well. Once the concrete had set, a long articulated truck drew up in the lane with the gleaming, freshly galvanized steelwork. The crane also arrived exactly on cue and the task of lifting them off the truck and lowering them onto the foundations began. It went like clockwork, and Ian's and my survey efforts with a sagging measuring-tape across the width of the burn proved to have been accurate. My seventy-six years-old father clambered about on the steelwork helping bolt the girders together, and he was happier than I'd seen him since my mother's death.

I was pleased that he had this opportunity to see Millfaid, having heard me enthuse over it for six months. But his stay with us at the flat in Blanefield was spoiled by his lapses into ill humour. While Song did her best to make him welcome and cook interesting meals for us when we returned tired and hungry from Millfaid, my father would look down his nose at what he considered to be this foreign food. Perhaps also he was smarting from Song's absence at my mother's funeral, or perhaps he was being cantankerous simply because he was no longer under my mother's emollient influence. At all events, it made it difficult for us to invite him again.

Over the following year, my father's visits to hospital got more frequent. I called on him whenever I was in the South and, fortunately, because of my meetings in London I was able to see him most weeks. Then, in April 1979, he was admitted to Harold Wood Hospital as an in-patient. His enormous energy and fighting spirit were leaving him and he was now seriously ill. The cancer had spread and he was in great pain. One morning I was preparing to leave home for my office when I received a telephone call. It was Dad. He was in tears and pleaded with me to do something to relieve his pain. It is heart-wrenching to hear from afar one's parent suffering, and I flew to London immediately and went straight to the hospital where I insisted on seeing his consultant. It was not that Dad was being neglected intentionally, but hospitals were not good at palliative care and the staff were over-worked and preoccupied with other matters. I refused to leave until they promised to be more attentive and to ensure his pain was controlled. Shortly after, he was transferred to Oldchurch Hospital and I continued to visit him as often as I could. He developed jaundice and went into a coma. On the 10th of May 1979 he died. In that last year since my mother's death, I had watched his will to live gradually ebb away, and now he too was buried at Hornchurch beside his wife and daughter. He was a complicated man, and often he was his own worst enemy. But I loved him in spite of his cussedness.

Soon after buying Millfaid I had engaged an architect. Foolishly, I had done this by looking in the yellow pages of the telephone directory and choosing one almost at random. We met on site and had a number of meetings at his office, but as the months passed I became increasingly disappointed by the ideas he was presenting – perhaps suitable for a suburban setting but not for a site as special as Millfaid. Eventually I had to tell him that we must go our separate ways. Then one day in the village shop I bumped into an acquaintance, Anne-Marie, whom we had met at a party. When she heard that I'd sacked the architect she suggested I should visit her own house, since it had recently been built and had been designed by an architect from nearby Gartocharn. It was a delightful place and I was not surprised when she showed me cuttings featuring it in the magazine 'House and Garden'. Although our budget would mean something much less grand at Millfaid, I called the architect, John Boys, that evening. After I had outlined our requirements and described the location of our ruin, he replied: "I've driven past that old cottage so many times, and have always thought how good it would be to restore it. Yes, I'd love to do it for you."

John was a delightful man and became a good friend. Song warmed to his wish always to consult her on any design issue, since, as he said: "It's going to be Song who spends the most time in the house." A session in his office, poring over his evolving plans, was like a session with one's psychiatrist, as he delved into our lifestyle, our tastes and priorities, and the compromises we were willing to make. What emerged was a single-storey, L-shaped house with traditional white harled walls and a roof of Ballachulish slate. Viewed from the lane about a hundred metres away, it would look as if it had sat there on its low hill for centuries.

Nearly ten years earlier I had been introduced by Ian Rolfe to the cottage on the shore of Loch Linnhe which he and a few friends held on a long lease. Now back in Britain and living only a two-hour drive away, Song and I continued to spend weekends there from time to time as guests of the Rolfes, but in 1978 the opportunity arose to buy a share in the cottage. On old maps it was shown simply as 'Back Settlement' but its Gaelic name was Leachnasgeir, meaning 'Rock of the Cormorant.' The story of Leachnasgeir is an interesting one.

A lecturer in geology at Glasgow University named Ken Shiels was leading his students on a geological field-trip when he came across what then were the remains of the cottage. It had no roof, its walls were crumbling, and small trees were growing through the floor. But he was attracted by its location on a grassy meadow backed by wooded hills, with views northwest across the loch to the mountains of Morvern and Ardgour. He investigated the cottage's ownership and was successful in obtaining a lease on the ruin at a rental of, I believe, £10 per year. He and his wife then set about the task of restoring it. At this point several Glasgow University colleagues, including Ian, come into the picture, as they willingly lent a hand in the restoration work. Since the cottage was over a mile from the nearest road the Shiels decided to buy a boat for taking them, their tools, and building materials along the coast from their parked car. But they hadn't reckoned on the rapid changes in weather which can happen in the Highlands. They set off one autumn day in what seemed safe conditions, but a squall blew up, swamped the boat, and Ken and his wife Wendy were drowned. They had no buoyancy aids and their bodies were never recovered.

With an emotional stake in the cottage project and as a memorial to their late colleague, Ian and three friends took over the lease and continued restoring it. By the late 'sixties it was complete, and although basic it had great charm and was a wonderful weekend retreat. Downstairs had been floored, a sink with piped water from a hillside burn had been fitted, there was a bottled-gas cooker, a wood-burning stove, and furniture made entirely from wooden fish-boxes washed up on the beach; upstairs was a sleeping loft. Then in 1978 the landlord decided he wished to sell the freehold of the cottage. The four lessees cast around for a fifth person to share the purchase price and in view of my familiarity with the place I was invited to contribute. For a figure of less than a thousand pounds I became a co-owner of Leachnasgeir and in the years that followed, Song and I had many happy times there, sometimes with friends and family and often with the Rolfes.

As Chief Geologist of the national oil company I was finding I had a high profile in the profession, and before long I was being invited to talk at oil-industry conferences and take part in delegations to overseas countries to meet my counterparts. In the spring of 1978 I attended a conference and geological field trip in Oklahoma, and in the summer I made my first visit to India to make contact with that country's two state oil companies. But my most interesting trip that year was to the Soviet Union, to take part in *'British Week in Novosibirsk.'* It was a motley bunch of people that boarded the British Airways plane at Heathrow. Our leader was Sir Fitzroy Maclean, a tall and imposing Scot with a wealth of stories from his time as a junior diplomat in Moscow through Stalin's purges and his friendship with Tito during his undercover work in Yugoslavia during the Second World War. The core of the party was made up of John Maddox the editor of 'Nature,' Michael Compton of the Tate Gallery, Hugh Conway the aircraft designer, a Scottish police piper called Joe Wilson, the poet Anthony Thwaite (with whom I found I was to share a hotel room), and myself representing the North Sea oil industry. Once at our destination I was due to give a lecture at the institute of science at Akademgorodok, and there would be British fashion shows, recitals by the Gabrielli Quartet, and screenings of the film 'The Railway Children' – an interesting choice considering we were supposed to be showing the modern face of Britain.

Somehow the well-known industrialist and former MP, Sir Robert Maxwell, had got wind of the expedition. Although he was not in the official party he had obtained a seat on our flight to Moscow and as soon as we were airborne he was on his feet, leaning over to talk with each of us in turn. He was proud of his knowledge of the Soviet Union and of his friendship with its leaders, to the point that his boasting became ridiculous and embarrassing. No one in the party was allowed to disembark without having thrust into his hand a signed copy of a photograph showing Robert Maxwell and Kruschev embracing each other.

We were entertained handsomely in Moscow and in Novosibirsk, but were never short of 'minders,' whom we assumed to be KGB officials. At the official receptions caviar was consumed in large quantities, and as the vodka flowed we toasted the undying friendship between our two countries. Only once did we experience hostility or suspiciousness of the kind I had found in the Soviet Union on my previous visit, in 1968. We were in the restaurant of the monstrous Hotel Ukraine in Moscow, dining and merry-making, when someone among us thought it would be a good idea to get Joe to give a recital on his bagpipes. Try as we might to involve the other diners in Joe's reels and strathspeys, the atmosphere became chilly and it was clear from the expressions of the hotel staff that such irregular behaviour was a step too far.

I stayed on for a few days longer in Moscow to meet officials of the plethora of government departments responsible for oil and gas. Robert Wade Geary, Minister at the British Embassy, hosted a lunch for me to meet a number of them, and afterwards took me along to meet Sir Curtis Keble who was British Ambassador at the time. In his office looking over the Moscow River to the Kremlin beyond, he greeted me in an exaggeratedly loud voice: "Of course, the whole place is bugged," he said, "and so you must realise that everything we say is being listened to and recorded." It was his way of alerting me and, at the same time, of playfully teasing the KGB.

The star of the trip had undoubtedly been our leader, Fitzroy Maclean. He was well-known from his writing, and spoke sufficient Russian to be able to deliver witty and charming after-dinner speeches. I was sorry that I saw him only infrequently back in Scotland before he died, sometimes bumping into him at the airport or at Creggans Inn, his family-run hotel on the shores of Loch Fyne.

I think it was in 1979 that I was asked to review the oil potential of the Falkland Islands. A meeting with Foreign Office and Department of Energy officials followed at which I learned something of the background to the request. A small amount of reconnaissance seismic data had been acquired around the islands and there was a growing wish in Government to know if there was any potential for finding petroleum there. The islands themselves are composed largely of volcanic rocks and have no potential, but offshore the prospects were less easy to determine. There was a blanket of sedimentary rocks overlying the volcanic 'basement' and although there was no direct evidence of the nature of these sediments some analogies could be made with offshore Argentina where, indeed, oil had been discovered. Our report concluded that there were some grounds to be encouraged, although a closer grid of seismic data would be required to pinpoint actual prospects worth drilling.

A year or two later I received an unexpected telephone call. The caller explained that he was the Commander of the Royal Naval ship '*Endurance,*' the survey and patrol vessel stationed in the Southwest Atlantic. Back in Britain on leave, he was trying to alert the Government to what he saw as the growing interest in the Falklands being shown by Argentina. He asked my opinion of the oil prospects and before he rang off he urged me to do anything I could to stir our own Government into taking more interest. I would have forgotten the call and the conversation had it not been for the Argentinian forces landing on the Falklands in 1982 and the war which followed.

With the Corporation's growing portfolio of acreage and responsibilities I needed to increase my staff numbers. At the junior end of the scale we joined the 'milk run' and were successful in attracting a healthy supply of young men and women graduates. But recruiting more senior exploration professionals was more challenging. I did manage to persuade some of my former BP colleagues to join us in Glasgow, notably John Nicholson whom I had recruited to BP in Australia some years earlier, and Colin Maclean who I appointed head of production geology, concerned with optimising the drilling on oil and gas fields once they have been discovered – both excellent men.

Our recruitment problems persisted and early in 1982 I agreed with Personnel Department that we should attack the Calgary market. Whereas the industry was booming in the UK, Canadian exploration was going through a down-turn. After advertising in the Calgary press a couple of us flew there and took a suite in one of the large hotels. We were not inundated by applicants but we did interview a number and made several offers which resulted in us filling one or two key positions in Glasgow. Before returning home I decided I would hire a car and take a look at the Rockies. It was April and snowy. I got to Lake Louise where the sun shone from a clear blue sky and I could not resist the opportunity of a day's skiing. I hired boots and skis and, still dressed in my business suit, I spent a wonderful day on the slopes – no doubt to the amusement of my fellow skiers in their stylish and colourful gear.

In my early days with BNOC I had decided that we needed a geological laboratory to handle the analytical work on rock samples, such as age-determination and source-rock geochemistry. I thought I knew exactly the right man to run the laboratory – my good friend and colleague over many years in BP, Ron Walters. I knew that he was unhappy to have been posted back to BP's research centre at Sunbury-on-Thames, and so I contacted him. Ron was enthusiastic and agreed to come to Glasgow to discuss the job. Although we had corresponded, I hadn't seen Ron for several years and so was unprepared for what I saw. After flying from London the previous night and staying in a Glasgow hotel, he arrived at my office for our agreed nine o'clock meeting. He was bleary eyed, his hands were shaking, and there was the smell of alcohol on his breath. Ron tried to explain it away by saying that he had been unable to sleep that night because of his excitement over the prospect of escaping his job and joining the Corporation. This, he said, had led to an attack of nerves and a bout of vomiting.

The meeting was a disaster and it was all too clear that I could not take him on. I wrote to him the next day explaining how disappointed I was and encouraging him to take himself in hand and sort out his problems. I still have his reply, hand-written in that elegant script that he had taught himself as a self-improvement project early in our friendship. His letter acknowledged that he was no longer the man he had been, putting it down to various domestic problems between him and his wife, Bernice. He was in denial over the real problem that had wrought such a change in him, his alcohol addiction. We never saw each other again. He left BP and was taken on by a consulting company to work as a rig geologist in Bangladesh. It was there that he suffered a liver disease and died in London in April 1985, a tragic end to a very fine man.

One significant difference from my BP experience was that in BNOC I moved among the most senior echelons of the Corporation's management. Lord Kearton had a reputation for ignoring managerial rank, and he would telephone me to ask about the geology of this or that oil or gas field. Then there were a number of committees that I was asked to sit on with various General Managers and Managing Directors. One of these, the Exploration and Appraisal Committee, met weekly under Kearton to discuss priorities and strategy. Alastair Morton was also on the committee and I came to admire his drive and his intellect, even though a meeting hardly went by without us getting into an argument. The reason we were so often at loggerheads was the fact that he had no experience of oil exploration and yet, because of his self-belief, he considered that it could be conducted like any other business. He could not come to terms with the way that exploration decisions are made: weighing very many factors, some of them unquantifiable, before concluding that, say, Prospect A should be ranked above Prospect B. Years later when he was making the Channel Tunnel a reality and was awarded a knighthood, I wrote to congratulate him; he was generous enough to reply that he had enjoyed the squabbles we had in BNOC, and that he had found me a worthy sparring partner.

On those months when board meetings were held in Glasgow the senior managers were invited to lunch with the directors. And so over the prawn cocktail I found myself talking with the likes of Clive Jenkins, the white-collar trades union man and champagne socialist, or with merchant bankers like Graham Hearne and Roy Dantzic. We would get visits from prominent politicians too. If one Labour Government minister could be said to have been the midwife of BNOC it was Tony Benn, demoted from Industry Secretary to the Energy Department by Prime Minister Wilson. I remember talking with him before lunch on one of his visits to Glasgow. It's a brave or foolish man who takes on Benn, and

I recall arrogantly questioning the wisdom of squeezing the oil companies too hard in his zeal to do the best for Britain.

One of the dominant themes of my presentations to the Exploration and Appraisal Committee was my belief that BNOC should not limit its remit to the United Kingdom but should explore overseas. My reasons were, first, that unless we diversified our areas of production we would be sentencing ourselves to an inevitable decline as UK production eventually declines; and second, to attract and retain first-rate professionals we needed to offer them wider horizons than simply the UK. Initially there was strong resistance to this proposal, particularly from those like Kearton who were driven less by the vision of building a major oil company and more by the imperative of exploiting the UK's resources for the benefit of the nation. But Kearton did see benefits in state oil companies co-operating, and in the autumn of 1978 a small group of us paid the first of several visits to Malaysia for discussions with Petronas. The state oil company wished to develop the capability to explore and produce oil as an operator, not just as a partner alongside operating companies such as Esso, and these discussions led to us seconding a team of personnel to Kuala Lumpur to help and advise them.

In 1979 Lord Kearton retired and the Government, unable to attract a top man from the oil industry, appointed Philip Shelbourne in his place as Chairman. Previously a lawyer and merchant banker, Shelbourne could not have been more different from Kearton. He was a City grandee, dapper, softly spoken, precise, and probably gay. He more readily saw the need to develop an international dimension and he approved my going to Los Angeles for discussions with the American oil company, Arco. In the UK, Arco had been courting BNOC as a way of acquiring the operatorship of blocks in the North Sea and the *quid pro quo* was that BNOC would be invited to join in some of its international ventures. Over the course of several days at Arco's head office I was shown a basket of projects. Some I immediately rejected as too high-risk, but two struck me as worth following up in greater detail. I returned home and recommended to the newly formed International Committee that we should investigate the possibility of joining Arco's ventures in Dubai and Indonesia. The Dubai acreage was onshore and interested me because, in my view, the previous holder of the acreage had wrongly interpreted the results of its drilling campaign there; if a further well were drilled at a different location it would have a good chance of making a discovery, I believed. The Indonesian acreage was a large offshore block called Kangean and lay in deep water north of the island of Bali. It had seen little or no drilling, but seismic surveys showed a variety of plays which, I felt, could be matured by further work into attractive prospects which would be worth drilling.

My recommendations were accepted, I carried out the technical 'due diligence', and BNOC signed-up as partners in the two ventures alongside Arco, the operator. The Dubai project moved fast and our first well, Margham No 1, made a major gas and condensate discovery. The Kangean project in Indonesia took longer but it too was successful and was a significant attraction for BP when, some years later, it took over what by then had become the privatised company, Britoil.

Philip Shelbourne's arrival prompted Alastair Morton to tender his resignation, as it was widely known that in their merchant banking pasts they had intensely disliked each other.

Morton's replacement was Malcolm Ford, an engineer previously with Shell and now nearing the end of his career. He arrived at BNOC early in 1980 and was given overall responsibility for UK exploration and production. He was to work alongside the Corporation's other Managing Director, Iain Clark. Clark was a Glaswegian who had worked his way up the local-government ladder to become Chief Executive of Shetland Island Council. The discovery of oil in the far north, and the need to bring it ashore on the Shetlands, had put the islands and its Chief Executive on the map and Clark had acquired a reputation as a tough negotiator. Clark was now made responsible for our growing international portfolio. It was around that time that my boss, Dick Fowle, decided to resign and I was promoted to General Manager to take his place.

While still Chief Geologist I received a letter from the Principal and Vice Chancellor of Strathclyde University, Sir Samuel Curran, inviting me to become Visiting Professor in the Department of Applied Geology. I had got to know the Head of Department, Donald Duff, and I was pleased to accept. I did give one or two lectures on petroleum geology but on the whole I was disappointed not to be drawn more fully into the department's affairs. I was also invited by Sir Keith Joseph, the minister responsible for education and science, to become a member of the Natural Environment Research Council (NERC), a quango whose purpose was, and still is, to advise the Government on matters related to the environmental sciences, including geology, and to apportion research funds made available by Government. Sir Hermann Bondi was Chairman, a likeable man, and other members included such worthies as Crispin Tickell and Lord Cranbrook. Ron Oxburgh, a geologist and later ennobled and appointed Chairman of Shell, was also a member and my memories of him are of his orderly mind and lucid utterances. On a more frivolous level, I remember he always wore short-sleeved shirts as if he was just back from a spell in the field.

I was dogged by a feeling of impotence at those NERC meetings, since it seemed to me as if major decisions had already been made by the secretariat in Swindon, and that we were being manipulated. Our role was therefore to be seen to have discussed the issue and approved NERC's proposals with no more than minimal alterations. Of much more interest was my membership of one of the committees reporting to Council, the Earth Sciences Committee, which dealt with research carried out in universities and the British Geological Survey (BGS). As the oldest geological survey in the world and an acknowledged centre of excellence, the BGS was highly regarded by the earth science community. NERC was the umbrella organization beneath which the BGS was intended to carry out its function. A tug of war went on between the two, and many of us fought hard to preserve some of the BGS's autonomy and its status, but NERC's central bureaucracy was a powerful and immovable force. My involvement led on to a closer association with the BGS and I was asked to join its Programme Board, a position I held and enjoyed until the early 'nineties.

This growth in the professional side of my career included a period on the Council of the Geological Society at a time when Professor Janet Watson was President, and a term on the Council of the Geological Society of Glasgow. I remember that the speaker at one of our Glasgow meetings was Algy Cluff whom I had got to know through my old friend David Robertson. His address that evening was titled "How to start your own oil company." Algy, a quiet and urbane ex-guardsman, had done exactly that and Cluff Oil succeeded in obtaining a share of the Buchan Oilfield in the North Sea. He put it all down to good timing and opined that it could not be repeated. But I wasn't so sure, and perhaps the seed of an idea was planted in my mind that night.

BNOC's newly appointed joint-managing director, Malcolm Ford, was now my line manager. It wrankled with him that I should have these outside professional interests which, he felt, took up too much of my time. But we fell out over other things too. He was of course an experienced oilman but all of his career had been on the production side, and as far as I know he had never before managed exploration. In his role as MD in charge of the UK offshore area (the United Kingdom Continental Shelf – UKCS as the Government called it) he charged me with advising him on the level of exploration drilling we needed to do to maintain our discovery rate and therefore maintain our level of oil production. I could not make him understand that it was impossible to answer the question; in fact it was not even a sensible question. On the UKCS the size of the oil discoveries was already decreasing – as my former boss in BP had pointed out, after the early flush of giant discoveries we were now "picking up the loose change." I argued that we should tackle it from the other direction: we should consider how much it was reasonable for us to spend on exploration drilling and estimate from that how much oil we could expect to discover, taking into account the drilling success-ratio on the UKCS, and the average size of discovery being made. We argued hotly from our different viewpoints and, whereas I admired Morton's contrary approach, I came to dislike Ford and his bullying manner.

Early in 1981 I received a call from a firm of London head-hunters. Their approach was oblique, enquiring if I knew of a person who may be interested in becoming managing director of a medium-sized exploration and production company. As he described the position my own interest was awakened; his timing could not have been better as I was becoming increasingly disenchanted with BNOC. He revealed that the company was Tricentrol, a company I knew well, and he agreed to send me further particulars. Talking it over with Song, we agreed that it would be worth looking into. The next step was a meeting in London with the group Chief Executive, Graham Hearne. Graham had been a non-executive director of BNOC and so I had met him from time to time. Now, in his office in the City he explained the nature of the job. The position was Managing Director of the UK business; (in addition Tricentrol had various interests abroad). As Graham described the company, talking about its "market capitalisation" and how it might raise further capital using "its paper", I realised just how ignorant I was of apparently everyday business terms. I concealed my ignorance as best I could and a few days later he wrote to offer me the job. I turned the offer down, not because it was an unattractive offer, but because when the moment came for me to decide, I realised that I simply could not bring myself to sell Millfaid and relocate to London.

We had moved in to Millfaid in March 1981. With its dozen acres, its small river, an oak wood, and views of Ben Lomond from the kitchen window, it was the kind of place I'd day-dreamed of when I was abroad, thumbing through advertisements in Country Life. Song too was happy. She had played a large part in the planning of the house and in its interior design. But her most important source of happiness, and mine, was the birth of our son Timothy a year earlier.

February 4th 1980 was a snowy day in Glasgow when Song went in to the Queen Mother's Hospital. Timothy was born at three minutes past six that evening and I was present throughout. It was the most moving experience of my life. He was a month premature and weighed only 2.74 kilograms. I gave him the nick-name Titch, which has stuck to this day.

Later in 1981 I heard again from the head-hunters. They had still not filled the Tricentrol position and this time I gave them the name of an old friend in BP who I thought might be interested. But the thought started growing in my mind that perhaps I had been too unimaginative when I turned down Graham's offer. Maybe a way could have been found to enable me to take the job and hang on to Millfaid. The thought buzzed in my mind for the rest of the year. Then, after a particularly frustrating day at the office in December I decided I would get in touch with the head-hunters again and, if the position was still unfilled, I would throw my hat back in the ring. The job was still open and they seemed pleased to hear from me. Again interviews with Graham were arranged. Tricentrol's Chairman and major shareholder, James Longcroft, was a tax exile who lived in Switzerland and so I journeyed to Geneva to meet him. Graham made his car and driver available and Song and I looked at various houses in Chelsea and Islington we thought might be suitable.

He made me an offer and took on board many of the demands I made for housing assistance in London. I had by now taken up about two months of his time and he had a right to think that I would not turn him down a second time. But I did. He must have thought me an idiot, but when the decision-point came I realised not only that I was unwilling to sell Millfaid, but that I wished to continue living there.

By now it was 1982, and while still in the final stages of those negotiations I was called into Malcolm Ford's office in Glasgow. "The Chairman tells me" he said, "that there's a rumour circulating in the City that you're resigning to take up a position with Tricentrol." I had to admit that there was some truth in this but, as yet, I had not made up my mind whether or not to leave. He said peevishly: "Well, you'd better decide soon and let me know." Shortly after that confrontation I was invited by the Chairman to dine with him at his London home as he wished to talk over the situation with me. A bachelor and *bon viveur,* his cook laid on an excellent dinner at his fine Georgian house on Aubert Park. We sat, the two of us, talking of pictures and of music, before he raised the matter of my possible departure. I outlined my grievances: the fact that exploration was regarded as a service to the international and the UK management teams – as if it were no more than a typing pool or a personnel department, instead of being central to the development of strategy; the fact that I had difficulty in working with both managing directors, Ford and Clark, whom I regarded as bullies in their different ways.

Whether Shelbourne thought my rejection of the Tricentrol offer was the result of our *tête â tête* I don't know. But I did notice a change in the perks which came my way. There were tickets to Wimbledon; he arranged that Song and I would be invited to a Buckingham Palace garden party that summer; and when I proposed at one of our International Committee meetings that I should attend a geological conference and field trip in China later in the year he over-ruled Ford's misgivings and heartily supported my going.

I decided to call on Arco in Dubai on my way to China. Our first well there had been a success and had confirmed my earlier interpretation of the geological structure – that it is an overthrust anticline – and now Margham No 2 was drilling. I discussed these findings and the planned seismic survey with our Arco partners, and I then flew on to Karachi. I had arranged to meet Iqbal Kadri, a senior geologist with the Pakistan state oil company. He was a charming man with perfect English and he laid on what turned out to be a delightful evening. Back at his home his wife, Mumtag, served a splendid Punjabi meal and afterwards his daughter, Sharma, and her friend danced for us. It was the kind of occasion one wished one could bottle, to open and savour at will later. I was up at four o'clock the next morning to catch the Air France flight to Beijing.

It was the sixtieth anniversary of the Geological Society of China and the conference was a way of saying to the outside world that the 'cultural revolution' was over and they were joining the community of nations again. In Beijing we were greeted in the Great Hall of the People by Vice Premier Wan Li and his Minister for Geology and Mineral Resources. I found myself seated next to one of the official interpreters, an attractive girl named Ji Hong. She explained that her original name was Lin-ay but that it was changed during the cultural revolution, the 'Hong' being Chinese for 'red'. I was able to converse with many Chinese academics over the following fortnight and almost without exception they told distressing stories of the so-called class-struggle and having been forced into manual labour by the infamous Red Guard in the dark days of the decade which had ended only six years previously. They explained that the policy was intended to remove "bourgeois" influences from professional workers, particularly their imagined tendency to have greater regard for their own specialized fields than for the goals of the party. But they were now able to look back without rancour on what they recognized as a phase of national madness.

After a couple of days sightseeing – the Great Wall, Ming tombs, the Forbidden City – we boarded a very comfortable sleeping-car train to Beidaihe on the Gulf of Bohai where the conference was to take place. It was a charming, quiet, seaside resort of clean beaches backed by pine woods. Scattered among the trees were holiday cottages which, from their cleanliness and comfort, were obviously for top people within the Party. As guests of the government these were to be our accommodation. I was allocated Number 134 and just a hundred metres away was Number 125 which had been the villa of Jian Qing, Mao's wife and a member of the famous 'Gang of Four.'

While the conference was interesting and I was able to meet delegates from all over the world – although oil company representatives were heavily outnumbered by academics – the field trip which followed was of even greater interest. The plan was to carry out a crossing of the Tibetan Plateau. Since the plateau's altitude of 3000 metres plus was a potential health hazard we all underwent a rigorous health check beforehand. It was carried out in a Beijing hospital and was without doubt the most comprehensive health check I had ever had. Some delegates were rejected, including my old friend Professor Derek Ager of Swansea University, but the rest of us boarded a Russian Ilyushin 18 for Chengdu, capital of Szechuan Province, and *en route* we were served powdery rice cakes and a drink made from a kind of hawthorn berry (*Crateagus pinatifida* it said on the bottle) by stern stewardesses in Mao trouser-suits.

We had been warned by some of the Chinese delegates of Szechuan's reputation for persistent rain, and were told the ancient proverb: 'In Szechuan, dogs bark in alarm whenever the sun appears!' Sure enough, we landed in a tropical downpour which had not stopped for three days; the countryside was under water and stooks of harvested rice stood forlornly in flooded fields. I was keen to explore the city and wandered the tree-lined back streets where vines climbed across the fronts of the small wooden houses, and one line of open-fronted shops seemed to be making nothing but bed springs; further along a man was stuffing kapok into mattresses using a strange vibrating contraption powered by a bamboo bow attached to his waist and projecting over his head. In the early morning the night-soil men appeared with their long hand-cart tankers and bailing-buckets on ten-feet long poles, the entire kit made of bamboo.

For the first twenty-four hours in Lhasa most of us suffered splitting headaches from the altitude, but when we had recovered we spent the next two days being shown the sights: the Potala Palace, Drepung Monastry and north to the Sera Monastry. Aged monks in dirty yellow robes shuffled through the cobbled alleys and courtyards, disappearing through carved wooden doorways into dark chambers where ancient wall hangings and carved idols could just be seen in the glimmering light from yak-butter lamps. The contrast with the festooned power cables and grim industrial buildings crowding the 'Number one Receiving

Hotel' was stark. That evening I chose to skip the film show and instead walked into the heart of the old city. There was no sign here of Chinese repression. At the entrance to the Jokhang Monastry Tibetans were bowing and prostrating themselves while others paraded slowly round the town in a clockwise procession, turning their prayer wheels as they went. They seemed uniformly dirty, as if their persons and the draped heavy woollen clothes they wore were ingrained with years of yak-butter lamp smoke. Some of the men wore a dagger at their waist, and some were prostrating themselves as they circled the city centre, measuring their length like inchworms.

We set off west in a fleet of Chinese jeeps on 10th September. Our traverse followed the Yarlung-Zangbo Suture, the line along which the crustal plate of India collided with that of Asia, and so there was much of geological interest to see. By way of the Gambala Pass – at 4800 metres the highest I'd then crossed – to Xigatze; to Dala; and over the Gyacola Pass at an even higher 5250 metres. Some on our party, particularly those who smoked, suffered badly from the altitude and had to be given oxygen dispensed from what looked like inflated rubber mattresses. On we travelled to the town of Tingri with its hilltop monastry sacked in the cultural revolution, then north around the foot of Mount Everest and into the magnificent gorge which drops through luxuriant rhododendron and cedar forests into Nepal. It had taken six days to reach the border, staying each night in scruffy rest-houses. We were now at Zhamu, an important border crossing, and lines of barefoot Nepalese porters came and went, straining beneath immense backpacks slung from bands around their foreheads. Our hotel was a series of tin huts hugging the side of the gorge. At only 2500 metres, Mike Brookfield my room-mate commented as he prepared for bed: "Well, at least taking your shirt off here doesn't make you pant for breath!"

When the Conservatives came to power in 1979 it soon became clear that a state oil company was not part of their thinking. It was decided that BNOC should be privatised. A scheme was worked out and on 1st August 1982 all of the Corporation's exploration and production activities were transferred to a new subsidiary called Britoil Limited. A letter was sent to all staff advising us that henceforth we would be working for Britoil, not BNOC. It was intended to assign all of the shares in Britoil to the Secretary of State later in the autumn, prior to flotation. Preparation of the offer document meant that I was exposed to a range of commercial matters and terminology that were new to me. With memories of the commercial ignorance I'd displayed to Tricentrol still fresh in my mind, I decided something needed to be done to rectify that yawning gap in my knowledge, and in early 1983 I enrolled on a two-week advanced course at Manchester Business School. I was taught a lot of management theory of doubtful usefulness, but the lectures on finance meant that by the end of the course I could at least read and understand a company's balance sheet. Back in Glasgow, I decided that my exploration staff should also have these basic financial skills. After all, one day they may wish to run their own companies and they should have the necessary commercial know-how. And so I arranged a series of in-house courses to be given by teachers from Manchester, and the initiative was well supported and appreciated.

Meanwhile, my frustrations with my working arrangements and my dislike for Ford and Clark had prompted me into secret discussions with two one-time colleagues who were working together to build a consultancy business: David Robertson, my old friend from Singapore, and Clive Hardcastle who had been BP's commercial man in New York during my time in Alaska. These talks came to naught but the idea was growing in my mind that I should leave Britoil and carve a new and independent niche for myself.

My travels on Britoil business continued. I was in Paris one week, Norway the next. Song and I travelled with baby Timothy to Bressanone in the Dolomites where I presented a paper at a NATO-sponsored conference on the geology of the northeast Atlantic. And in September 1983 I made my third visit to India, to present a joint-authored paper at the World Energy Conference in Delhi. (As well as my visit there on business in 1978 I had made a private trip with a friend in 1981, when I criss-crossed India by train in the course of a three-week vacation). The great and the good were at the conference, including our former chairman Frank Kearton and, of course, his replacement Philip Shelbourne. I remember little about the conference proceedings but I do remember that while I was standing in a queue in the hotel coffee shop one lunch time, a young woman who was also standing in line started talking to me. Her name was Eliza, she said. She was Italian, and an attractive girl. She was there again in the evening and we had dinner together, during which she talked of her life " ...with the children." I was slow to grasp what she was talking about, wondering if perhaps she worked with a childrens' charity or at an orphanage. But by the time I left Delhi I had discovered what she meant. She was a member of a community which was a branch of the worldwide "Children of God," an evangelical movement which uses sexual favours to lure people into its fold. Poor Eliza! I must have been a disappointment to her.

Life at Millfaid was good. In the spring of 1982 we paid £1000 for a pair of Highland cows with calves at foot, and in the following spring our small herd swelled to six with the birth of two more calves. In the summer months we would send off the cows and mature heifers to run with the bull and nine months later more calves would be born, looking like fluffy teddy bears. We also acquired a couple of pairs of geese which soon multiplied to produce a flock. They spent the nights on an island in the burn, finding that this gave them protection from prowling foxes. But we had some severe winters in the early 'eighties, and when the burn froze the geese became easy prey.

We kept a few hives of bees too, and extracted enough honey to give to friends and family, until I found one year that when stung by a rogue bee I developed an allergic reaction. It was time, I realised, to move on to another form of livestock.

A local farmer let me have a few sheep, a motley lot including some Jacob and some Black-faced ewes. There was a ram among them and so the flock soon grew. But sheep – at least in the wet conditions of the West of Scotland – seem to have a death wish. They either get attacked by blow-flies, or by internal parasites, or their feet rot, and so they need frequent attention. Without a sheepdog it could take a while to round them up and get them penned, but that was the easy part. Once in the pen the fun started. I would get among them and, choosing my sheep, would grab it and try to force it to the ground. Rams and old ewes can grow to a fair size and they could put up a good fight. Getting them onto their backs so that I could sit on them and subdue their flailing legs was like doing three rounds with a professional wrestler.

One amusing distinction between a farmer and an agriculturalist that I have heard is that a farmer makes money in the countryside to spend in the city, whereas an agriculturalist makes money in the city to spend in the countryside. I certainly fell into the latter category. But even if it was not profitable it was spiritually rewarding. Just living at Millfaid was a pleasure, whether I was tending our animals, felling a tree or digging a ditch, or simply doing the rounds of our tiny estate. Driving home from Glasgow after a day at the office, the Drymen road climbs the gentle incline of the Kilpatrick Hills to a low pass called the Queen's View, and suddenly the Highlands and the bowl of Loch Lomond are spread below. Rain or shine, that vista never failed to lift my spirits, and the knowledge that our house and our patch of land were down there among the forests and fields made the feeling all the stronger.

I decided in 1984 that I would leave the now-privatised Britoil. I had been with the company (including its forerunner BNOC) for seven years and although I had benefited from the experience it had not been altogether enjoyable. I had a fine team of people in exploration department, but my relations with those above me were intolerable. In particular the two managing directors to whom I reported were the most unpleasant colleagues I had worked with in my twenty seven years. Ford was a straightforward bully, abusing his seniority to shout and swear at his subordinates. Clark was a more complex person. Quietly spoken, scheming and a born-again Christian, he would ask a subordinate into his office and then seek to crush him, using assertions, accusations, sanctimony and sarcasm.

But on the positive side, my commercial awareness was now infinitely greater than when I had left BP. I had become well-known in the upstream oil industry and in the profession, and had a wide circle of contacts. And because of the breadth of the company's involvements across the offshore UK my knowledge of its geology was considerable. Even from Malcolm Ford I picked up one useful habit. He was never without his spiral-bound notebook, in which he jotted highlights of meetings, names, addresses, and anything he felt he might wish to refer to in the future. I started doing the same that year and am now on number 82.

I decided that I would become an exploration consultant, and hoped that my long list of contacts would bring in some requests for advice or geological assistance. I talked it over with Song and, although she recognized the risks associated with complete independence, she was very supportive and encouraged me to take the plunge. On 6th September I submitted my formal letter of resignation and started my new career on 1st October.

Chapter Fifteen

We converted a spare room at Millfaid into my office. I bought the latest IBM electric typewriter (with a useful device for erasing typing errors), a leather-topped 'partner's' desk, a filing cabinet and a telephone answering machine. And then I started casting my net hoping to land some clients.

On an earlier flight to Los Angeles for Britoil I had found myself seated next to Ewan Brown, a partner with the Edinburgh merchant bank, Noble Grossart. In the course of the flight he'd told me about Pict Petroleum, the small oil exploration company Noble Grossart had established, and we'd talked about the portfolio of North Sea licence interests it had built up. Recalling that conversation, I called Ewan, explaining that I had now left Britoil and was setting out as a consultant. He was interested and invited me over to Noble Grossart's Queen Street office. This brought me my first client, and within a couple of weeks of leaving Britoil I was representing Pict at meetings in London and Aberdeen with its joint-venture partners. Ewan would accompany me to some of these meetings and I developed a strong admiration for his sharp intellect, his unpretentiousness and his quiet negotiating skills. Monsanto was one of Pict's operating partners, and on their joint North Sea block 15/21 it had made two oil discoveries. Although still only at the appraisal stage, these two discoveries, named Ivanhoe and Rob Roy, were the jewels in the crown of Pict's portfolio of interests. They would transform the company's future revenue and its value, and much of the time I spent working on Pict's behalf was seeking to ensure that the operator would bring these discoveries into production as expeditiously and efficiently as possible.

The American company, Amerada Hess, was another that responded to my call. I remember staying for several weeks in their company town-house near Holland Park through the early winter of 1984-85, commuting daily to their offices off Tottenham Court Road to work on seismic data from west of the Shetlands which led on to Amerada's successful application for acreage in that so-called 'frontier' area. A less satisfactory result of that association was the discovery that some companies, including Amerada, enhanced their cashflow by the practice of late payment of invoices. I recall that I had to wait many weeks to receive the fees I was due, and finally resorted to asking Sam Laidlaw, Amerada's Managing Director, to instruct his finance department to pay up.

Another client was the Japanese company Sumitomo. I had come to know one of their London staff, Max Inanaga, and had met their Managing Director, Hirota, and Exploration Manager, Mizutani, when they had visited Britoil some months before. I was now asked to assist them in finding, and then buying, a North Sea oilfield interest. The field we settled on was the Balmoral Field and it took several months to negotiate the purchase, but when the interest was finally secured it gave Sumitomo a solid foundation on which to build a North Sea exploration and production business. By now it was 1985 and we'd established a close social as well as working relationship.

My consulting career was progressing well and in addition to my two main clients a number of small jobs came my way. Charging £400 per day my early anxieties over whether we could survive financially were now behind me. It must have been some time early in 1985 that I was in conversation with Iain Harrison, a neighbour and good friend whom we'd met soon after acquiring Millfaid; in fact, not many months earlier our families had enjoyed a skiing holiday together at Saas Fe. I remember, we were in the garden at Millfaid and Iain was asking how my new career was developing.

"It's going pretty well, thanks Iain, but you know, what I feel I'd really like to do is set up my own small oil exploration company." He considered this for a while.

"Roughly how much do you think you'd need to start a company like that?" he asked. I was not expecting the question, and plucking a figure out of the air, I replied.

"Oh, I should think about half a million would get it up and running."

And that is how the next, and most exhilarating, phase of my career came about.

In the months since leaving Britoil a number of small oil companies had come across my path. Some were partners of Pict's and I'd got to know their management teams around the table at different committee meetings. Others were companies with little or no technical staff which had contacted me to carry out some work for them or invited me to join their staff. One person in particular, Michael Seymour, who had established and was now Managing Director of a small company called Teredo was in my mind when I mentioned to Iain Harrison that I'd like to set up my own company. Michael had been a geologist with a major oil company and I felt that if he could do it, then so could I.

As for Iain, in the 'fifties he had founded and now ran a family ship-owning and ship-management company in Glasgow called Harrisons (Clyde) Limited. A patrician with a love and deep knowledge of music and the arts, he was also a generous and charming person with a wide circle of friends. In the thirty-odd years since he'd established Harrisons, shipping had made him a wealthy man, and yet such was his modesty that he would never have dreamt of displaying his wealth. His company, he told me, was doing quite well and his board had been discussing the possibility of diversifying. It was against that background that, following our conversation in the garden at Millfaid, he proposed to his fellow directors that they should make an investment in a start-up oil company.

In the days that followed I visited Harrisons' office and presented my ideas to their board. A round of onshore UK exploration licencing was being planned by the Department of Energy, and my proposal was that we should assemble a group of like-minded companies to apply for several of these blocks of acreage. The fact that we would be a British company would count in our favour with the Department, and we ought to be able to persuade them that we have the necessary technical skills. If we were awarded some blocks, our strategy would be to carry out seismic surveys to identify any prospects that may be present, and so having increased the value of the acreage we would then invite a bigger oil company to farm-in, paying the cost of drilling an exploration well to earn its interest. Among the major North Sea oil companies there was a healthy appetite for drilling onshore the United Kingdom, where at that time the cost could be offset against their offshore Petroleum Revenue Tax liabilities. My job of persuading Iain's colleagues was made easier by the fact that one of the company's subsidiaries operated supply-ships for North Sea oil companies, and so they were already involved in the service sector of the oil industry. It was agreed that Harrisons would invest half a million pounds in our new company.

We needed a name for this new company, of course, and Iain and I bombarded each other by telephone with our latest ideas. The suggestion 'Croft' came from Iain. Croft House, I knew, was the name of Iain's childhood home, and since I lived near the village of Croftamie it immediately appealed to me; moreover, the name had a good Scottish ring, and it would appear close to the top of any alphabetical list of companies. So we settled on the name Croft Exploration Limited. We agreed that Iain would be Chairman and I would be Managing Director.

While these developments were taking place, my consulting was continuing, in particular for Pict. I explained to Iain that I wished to continue with those arrangements, at least in the short term, and I suggested that I should set up a new company which would employ me and the other staff we would need, and that this consulting company would then sell its services to Croft and Pict. The name I chose for it was Liberty Exploration Company Limited, partly in memory of my old school and partly as a celebration of my independence. I decided that Liberty rather than me personally should be my vehicle for holding shares in Croft.

The capital structure of Croft that we agreed gives some idea of the fair-minded way in which Harrisons approached things. We agreed that the largest part of their investment (£499,000) should be in the form of redeemable preference shares – in other words a sort of loan – and that the remaining £1000 of capital would be in the form of ordinary shares which Harrisons and Liberty would subscribe for equally. The result was a robust half-million pound company over which we each had equal influence and voting rights.

The year 1985 was a hectic one. As well as continuing with my consulting work we kept the lawyers busy. Croft Exploration Limited was incorporated on 12th March. I then had to negotiate a new service agreement between Liberty and Pict (represented by Ewan Brown), and another service contract with Croft (represented by Harrisons directors). Separate agreements had to be drawn up between Pict and Croft to regulate how costs would be split and new opportunities would be shared, and another between me and my new employer, Liberty. At the same time I began to recruit into Liberty a small team of employees, starting with a personal assistant-cum-office manager and a geophysicist. I was fortunate to find a young Ulsterman named Jed Armstrong to fill the latter role. A quiet man with a sardonic sense of humour, he had been with the British Geological Survey and, as I was to discover over the following years, he could turn his hand to every aspect of geophysics, from planning a seismic survey to seismic-data processing to seismic interpretation – a range of skills that would normally be spread across an entire geophysical department!

Liberty took a lease on a top-floor suite of offices on Woodside Terrace, a Victorian stone building typical of Glasgow, and conveniently close to Harrisons. As these administrative matters progressed I began the long job of contacting other oil companies to secure places for Pict and Croft in bidding-groups. In those days there was a plethora of small and medium-sized companies with names now largely forgotten:

Charterhouse, run by another ex-BP man named Craven-Walker; Bula, run by a nimble-footed Irish accountant; Blackland, whose MD was a former Mayfair tailor and the son-in-law of John Betjeman; Deepwood, which was the oil-exploration arm of a quarrying business owned by Albert Rockarch; Fishermans, which had been formed by a group of Moray Firth fishermen; and many others, some credible and some less so. We agreed to join a group led by Michael Seymour's Teredo to apply for blocks in Central England, and another group led by Mobil which would bid for blocks in the Cheshire Basin. To have been invited into a group led by a major oil company was satisfying and seemed to confirm that we were moving in the right direction toward becoming a small oil company that would be taken seriously. All we need now, we thought, is to be awarded one of those blocks!

All of this activity did not prevent me taking the family to enjoy a short break in the sunshine. In March, Song and I with Timothy, now five years old, took a plane to Cairo. It was not our style to plan expeditions like this in advance, but rather to make our plans as events unfolded. It worked out well, and we did all of the things one associates with a trip to Egypt – the *souks* and back streets of Cairo; the pyramids; an overnight sleeper train to Luxor where we found a cruiser to take us up the Nile to Aswan; and a hired-car and driver to take us to Hurghada where we snorkelled in the warm waters of the Red Sea.

At Millfaid there was no shortage of jobs to keep me occupied in my spare time. I built a small stone shed – which became known as 'the bothy' – in the back garden; there were fences to repair; ditches had to be cleared annually of their debris; trees were planted and others felled. It was physical activity which complemented the mental exertion of building our oil business, and I revelled in it. Another short diversion from budget meetings and partner negotiations was a return to my childhood haunts in the Welsh Borders, where I decided I would undertake a trek. I'd done a similar trip the previous year, starting at Ludlow and reaching Abergavenny, but this time I would tackle a more northern stretch of the border. Song and Timothy drove me to Shrewsbury. After a night at the Lion on Wyle Cop I left the car in the hotel's car park, put Song and Timothy on a train back to Glasgow, and took a train myself to the village of Chirk, which was to be my starting point. An extract from my diary for Tuesday, 1st October brings back memories of that small adventure; it was my second day:

"Paid the bill (far too much, at £35, albeit including dinner and wine) and gave curly-headed Sharon a £2 tip, before leaving the Golden Pheasant Hotel at 9.30.

Took thirty minutes of hard trudging to reach the top of the plateau, then set off south in dry, hazy, sunny weather. Along moorland track, past forestry operations, until descending into a dingle and then hilly and liberally wooded farm country continued to Llansilin.

At Llansilin Inn, got a couple of half-pints down in the first ten minutes and then had a ploughman's lunch sitting out on the forecourt in the sunshine. Pleasant young skinny proprietor. (B&B here is £6.50, and I'd guess excellent value; c.f. the Golden Pheasant).

Continued south through very pleasant hilly farmland with abundant woods and trees. Sometimes along lanes and sometimes following (on the map, if not always apparent on the ground) footpaths across fields.

Passed an impressive motte and bailey about 1 _ miles outh of Llansilin, with a lake in the lower moat with hundreds of 'wild duck;' large flocks flew off at my approach. Presumably they're bred for shooting, as are the masses of pheasants around here which are almost possible to catch by hand. Shooting these could hardly be described as sport!

Lay on hillside, took off boots, and dozed for thirty minutes, overlooking small stream and wooded hillside. Then walked on to find store at Pen-y-Bont. Picked damsons in its back garden, overhanging the field, before going in and buying four postcards – pen and ink drawings of the store. Sat outside and wrote cards and ate damsons. Bought bar of fruit and nut chocolate, and continued south.

Damson trees are fairly common in gardens around this part of Wales and are just right now – ripe enough, but not so soft as to squelch in the pocket. Blackberries also excellent. Picked a large cooking apple too which hung over the road from one country garden.

Arrived at Llansantfraid (pron. Llan sant fried) late afternoon. Tried Lion Inn, which looked attractive in a clothing of red Virginia Creeper. Pleasant bearded landlord regretted that the decorators were in all the rooms, but suggested Glanvyrnwy Farm, through the village and across the river.

Mrs Jones was in the garden, hanging out the washing when I arrived. Nice, grey-haired granny who joked she had no spare rooms – then asked where I would park my car! Given pleasant clean double room in the front of house. Had hot bath and changed, then, about 7.00, sat down alone to massive dinner of soup, eggs, bacon, etc. Mrs Jones insisted I try one of her baked potatoes (if not two); certainly very good. I asked about them – 'desiree' apparently, cooked in the microwave. Then blackberry and apple crumble with ice cream. Felt very full!

Walked into village to call home. Song out, so left a message with Iain Harrison. Song called back later. Watched TV then read in bed, feeling very satisfied with life"

I walked on south over the next several days, through Montgomery, Bishops Castle, Hopesay and Leintwardine, following in places the line of Offa's Dyke, and spending the nights wherever I had reached by evening, and revelling in this unspoilt stretch of country. My last night was spent at the Talbot Hotel in Leominster from where I was able to take a train back to Shrewsbury. I lingered there, rediscovering its ancient charms and liking it more and more, before picking up my car for the drive back to Millfaid.

We spent that Christmas and New Year holiday in Thailand, staying for a few days with Song's sister, Nai, at the impressive new house at Hua Mark which had been designed by her architect boyfriend Jarin. I wanted to revisit some of the places where I'd worked fifteen years earlier and so we drove southeast to Chanthaburi, the centre of Thailand's gemstone industry. The small town I remembered was now big and sprawling, but the saddest part of that excursion was the extent to which the forests of this region had given way to agriculture. Where once we had inched our way along forest tracks there was now a network of tarmac roads serving the wide prairies of sugar-cane and other crops. The only reminders of the past forests were gaunt skeletons of burned trees dotted across the landscape.

In early 1986 our applications with Mobil and Teredo were successful and we were awarded licences to explore in the Elworth area of the Cheshire Basin and near Keighley in the Pennines. This is what we had set out to achieve and it was gratifying that things were proceeding according to plan. But this success meant that it was no longer reasonable for me to offer my services as a consultant. My relations with Sumitomo remained excellent, and to show their appreciation of the help I'd given them in purchasing an interest in the Balmoral Oilfield I was invited to visit Japan as Sumitomo's guest. I was, of course, pleased to accept, although they were slightly taken aback when I said that I should like my wife to accompany me. We agreed that provided I pay for Song's air ticket, Sumitomo would pay all of our travel and accommodation expenses in Japan.

Leaving Timothy at Millfaid in the care of two experienced childminders and Mary Gibson, a childrens' nurse and family friend, Song and I flew by way of Anchorage to Tokyo in early March 1986. We were treated royally by Sumitomo. In Tokyo we were wined and dined, and one evening while Song went to the Kabuki theatre I was introduced to

Ginza's nightlife, including the Kaji Club with its friendly English-speaking hostesses, apparently a regular haunt of certain of our Sumitomo hosts.

We took the train to Hakone Yumoto, a hot-spring resort in a wooded valley where we had our first experience of a *ryokan,* the traditional Japanese inn where the *futon* is prepared each night on a *tatami* floor and meals are prepared exquisitely on a low table in one's room. Other notes in my diary tell of the Tawaraya *ryokan*, and the gardens and temples in Kyoto; the *shinkansen* to the canal-side honeymoon town of Kurashiki; the Peace Park, Glover House, and a museum of Imari and Arita ceramics in Nagasaki.

It was at one of these towns, I think Hakone Yumoto, that we had our only language difficulty of the trip. After a day exploring the small town we returned to our *ryokan* and I went to the reception desk and asked if they had a telex facility, as there was a matter on which I needed to contact my secretary in Glasgow. The girl behind the desk looked blankly at me and I repeated my request, this time with gestures. She still failed to understand. As if in a charade, I pounded an imaginary typewriter with my fingers drumming, and kept saying the word 'telex.' Still unable to make myself understood I continued miming with my hands.

"I want to send a telex; you know, a telex? A message? A message? I want to send a message?" At last her face lit up. She seemed to understand.

"*Ah so deska!*" she said, exultantly, "what time you want?" Puzzled, I asked:

"Now? Can I send it now?"

She knew the word 'now' but regretted that this was not possible.

"Sewen o'crock?" she suggested. Frustrated, I gave up the struggle, settled for sending my telex at seven o'clock that evening, and joined Song in our room. At the appointed hour there was a knock at our door. I opened it to find, not the receptionist, but a smiling elderly woman in a white gown and carrying a basket of towels and other paraphernalia. That evening Song enjoyed a long and relaxing massage, and I resigned myself to not sending my telex message.

Our Japanese journey over, Song hurried back to Glasgow to satisfy herself that Timothy was well and happy, and I flew to the USA. Over the course of the next week I covered a lot of ground. I had never seen the Grand Canyon and so I hired a car in Las Vegas and set off east across the Hoover Dam and up the Colorado River. No one could argue over its magnificence, but standing there on the canyon's south rim, gazing at its vastness and imagining the hundreds of millions of years represented by the strata that form its walls, I soon grew tired of overhearing the inane comments of other tourists: "There sure is a hulluva lot of rock down there;" or "Only God could have created something like this." I drove on, sleeping for a few hours in a motel at the west end of Lake Powell, and then on to Zion National Park to see the sun rise on the towering pink and brown cliffs of Navajo Sandstone, while in the valley bottoms the first spring buds on the aspens were opening. By late morning I was back in Vegas, dropping off my Avis car and boarding a flight to Denver to stay with old university friend Mike Holmes who had settled there and carved for himself a flourishing oil consultancy.

In Dallas I called on Bruce Robson who looked after his family's shareholding in a small Texas oil company which I thought might provide Croft with an *entrée* into some US exploration opportunities. He drove me north to their office by Lake Texoma on the Red River. Although I decided against pursuing that idea, I learned much on that trip about America's small-scale approach to drilling and production. While major oil companies with household names are involved, there are a myriad small companies and individuals who also have a stake, perhaps by owning a drilling rig or having a well on their land which pays them a royalty. Some grow rich as a result, but others just make a living, and it is a fallacy to believe that if an American owns an oil well he is therefore probably a millionaire. It is more accurate to think of those 'oilmen' as we might think of a small-town builder or someone who runs a corner shop in Britain.

Before arriving back in Scotland from this whirlwind tour of Japan and the USA, I journeyed to Charleston in West Virginia to call on my friend, Ji Hong, the interpreter I'd met and got to know in China on my 1982 trip, and who by now had settled in America. It had been a memorable trip, and only possible through Sumitomo's generosity. And I'm pleased to say that two of those Sumitomo managers, Mizutani and Inanaga, are still among our close friends.

Throughout 1986 I had been a member of the committee organizing the next major conference on the petroleum geology of North West Europe. These conferences were held in London every few years, and this one, in October, was to be the Third Conference and was to be held at the Barbican Centre. But early that year, the countries of OPEC embarked on a strategy designed to increase their market share, and the resulting price collapse sent many oil companies into shock. We seriously considered cancelling the conference, fearing that belt-tightening in the companies would reduce their support. But we went ahead and it fell on me to be the final speaker, charged with reviewing the conference highlights. I remember likening the oil-price collapse that year to "Big Bang" in the City; but I was able to reassure the audience that although we had considered possibly cancelling the conference it had turned out to be a major success, with over 1200 delegates, and 120 technical papers.

For the time being Croft was not directly affected by the oil-price collapse – although indirectly we were, as I shall point out later. With cash in the bank, Croft was receiving a useful amount of bank-interest. But under Inland Revenue rules, we would have to pay Corporation Tax on that interest unless we could demonstrate that we were a *bona fide* oil company. Their criterion for us to qualify was that as well as exploring for petroleum, we must also be a producer of oil or gas. We obviously could not wait until we had discovered petroleum in one of our future exploration wells, and so it became necessary to purchase a producing interest from another party. This was one of the instances when our close relations with Pict benefited us. They had some small percentage interests in a number of oil and gas wells in Louisiana and Texas, each well pumping just a few barrels of oil or flowing a few thousand cubic feet of gas per day – so-called 'stripper production' of the kind I had visited in Texas that spring. With their North Sea fields Ivanhoe and Rob Roy now under development and due to come on stream soon, Pict no longer needed these USA interests and agreed to sell them to Croft. It was a piddling amount of oil and gas, but still it was another building-block toward what we were trying to achieve.

It was the following year that we were invited to participate in a well about to be drilled by a small independent company in Oklahoma. Confident by now that I knew something about the uniquely complicated American domestic oil and gas business, I agreed that Croft should buy a small interest in the well. And – demonstrating that even a simple geologist can be a part-owner of a well – I bought a five percent stake in the name of Liberty. The well discovered gas and remained in production for many years, bringing us a small monthly cheque; which in Liberty's case was not much more than enough to pay for the office Christmas party.

Song and I had become good friends with a family, the Hamiltons, who lived south of the Clyde, in Kilbarchan. Gordon was a GP, his wife Sri was Thai, and their daughter Jenny was the same age as Timothy. Gordon was the proud owner of a yacht which he kept on a mooring at Gourock, and in the summer of 1986 he and some friends had embarked on a cruise up the west coast of Scotland. Having got to Gairloch, north of Skye, his crew had to return to work in Glasgow, and Gordon asked me if I would join him and another friend to help sail his boat back to the Clyde. It was June, and I agreed enthusiastically as it had been years since I'd sailed.

Song drove me to Badachro, a small cluster of houses and a pub beside Gairloch, with the atmosphere and charm of a Cornish fishing village. After lunching in warm sunshine at the pub, Gordon rowed me out to Trebor, riding at anchor in the bay. A timber-built boat which had sailed the Clyde since she was built at the beginning of the century, there was just enough room in the saloon for us to spread our sleeping bags and for a gas cooker and sink – and for Gordon's collie which was happy to curl up in the forepeak beside the dripping anchor chain. The toilet arrangements were of the 'bucket-and-chuckit' variety.

We got to the Crowlin Isles that evening, entering the narrow slit of water which separates the two halves of land, and anchoring in a sheltered pool between heather-clad hills which seemed as though we were the first ever to have been there. Here is an extract from my diary:

"Fortified by bread, cheese and salami, with hot coffee and Bovril, we turned-in about 10.30, Robert in the for'd starboard berth, Gordon in the starboard and me in the port berth. Pretty stuffy, as Gordon wanted the hatch door virtually closed and the roof light was battened down. Slept fitfully because of a low step beneath my feet.

Monday, 30th June.
Woken by Gordon's alarm at 1.00 a.m.. Up promptly to catch the tide, started engine and away east of the Crowlins on 160°. Light enough to see outline of the islands, and with Gordon's compass light, steering was easy. Made four knots with engine. No wind, no clouds and not too cool. Manoeuevred through the different navigation lights and street lights of the Kyle of Lochalsh – easy provided you study the red and green navigation lights carefully – helped by the south-flowing falling tide. Interesting swirling overfalls in Khyle Rhea, then into the Sound of Sleat where I took another couple of hours of sleep just as the sun was coming up.

Woke about 8.0 a.m. to find we were still motoring south in the Sound of Sleat, with the Mallaig ferry crossing behind us. Superb warm and calm weather. Motored on SW while Gordon made a fry-up of eggs, bacon and onions. Got to Eigg about eleven and dropped anchor in the channel between Castle Island and the mainland of Eigg. Rowed ashore, still in a flat calm.

Delightful place, with a tearoom by the jetty, stacked high with home-made cakes, scones and pizza (awaiting pleasure boats with tourists from Arisaig about 12.30). Had lunch there and tea (two cups, since I'm missing tea badly) followed by an excellent crap in the clean loo behind the café. Then set off to the top of the Sgurr of

Eigg (1291 ft), an easy climb but sticky under the hot sun. Robert rested half way up and waited for us as he'd had too much sun yesterday. Spectacular view from the top, down onto the harbour and Trebor bobbing at anchor.

Walked back down and then north till we found the island's shop. Bought tea bags – hoorah – and milk, as Gordon was becoming difficult over the three cartons of fresh milk we'd consumed so far. Cadged a lift in a red Landrover the two miles back to the peir. I asked the driver if it was necessary to licence the Landrover in such a place, since it was the only vehicle we'd seen. "No, I don't bother." he replied. "Isn't there a policeman on the island then?" I asked. "Yes. You're looking at him!"

More tea and cake at the teashop just before five, closing time, and arranged to have dinner there when they re-opened at 7.30. Rowed back to boat in a rising breeze, but still no clouds. Went ashore at 7.00 and had a good wash and brush-up in the loo behind the restaurant. Then dinner – mushrooms on toast, chicken salad, cheesecake – quite pleasant, with the cook, the husband (and the baby in a pram) eating there too.

Tuesday 1st July.
The wind got up in the night, and in the channel the effect of wind against tide gave us a bad beating – nearly thrown out of my bunk, and five-gallon drums crashing about, until we got up at 5.00 a.m. and re-anchored in quieter waters by the jetty.

Got away about 9.00 and sailed south on a broad tack past Muck and Ardnamuchan in sunny weather and wind, now moderated. The wind freshened and we were unable to land on Staffa in the late afternoon/early evening, but we sailed close and the basalt columns of Fingal's Cave looked dramatic, picked out by the shadows of the setting sun.

Wind continued freshening and we made about 7 knots across to the Sound of Iona and into the calm water of Bull Hole, off the SW tip of Mull ....."

We took the Crinan Canal route back to the Clyde and ended the voyage at Gourock, as planned. I enjoyed every minute of it. I had not imagined that cruising the West of Scotland could be such a delight, particularly the joy of entering a quiet anchorage on some uninhabited coast at the end of an energetic day, feeling as Captain Cook might have felt when he dropped anchor off some Pacific shore. I was hooked, and was now in the market for buying a boat of my own.

Back at the office I and my small team in Liberty continued to juggle the affairs of Pict and Croft. The UK licences newly-awarded to Croft – in Cheshire and the Pennines – were operated by Mobil and Teredo respectively, and although we were therefore not required to man the exploration operations, they did involve a lot of additional work. There were joint-operating agreements to negotiate among the partners, budgets and work-programmes to digest, to discuss and agree, and then, as the technical information obtained from those particular areas started to arrive in our office, we needed to interpret the data ourselves so that we could hold our own in the planning of further work. Our staff of only three was no longer sufficient. A young geologist was recruited, and a girl to help with the office jobs, along with a librarian to catalogue and care for the growing volumes of technical data – the life blood of any exploration venture. Our top-floor offices on Woodside Terrace were barely adequate, particularly as we started to hatch plans to evolve from a non-operator role to that of operator. The answer was a suite of offices in Fountain House, just a few hundred metres away on the corner of Sauchiehall Street, where as well as all the space we were likely to need for several years to come, we also enjoyed the luxury of several car-park spaces.

It was important to continue building up Croft's portfolio of exploration interests. Since, statistically, exploration drilling has only a ten to twenty percent chance of making a discovery, the wider we spread our net the greater our chance of success. While we continued studying Britain's onshore sedimentary basins, coming up with ideas for future ventures, we decided that Croft should take its first step offshore. Michael Seymour and I seemed birds of a feather and Croft was already a partner with his company, Teredo, in the Pennines and in a Northern Ireland block. Now we agreed that Croft and Pict should join a group led by Teredo applying in the Government's Tenth Round of offshore licencing. The group was successful and we were awarded two blocks in the Southern North Sea gas basin.

It is a paradox that when an exploration company achieves its goal of discovering oil or gas, before it can enjoy the fruits of that success it faces the problem of having to pay large sums of money to bring the discovery to production. Nowhere is the problem greater than offshore. Although Pict had grown considerably in value since Noble Grossart had set it up, it faced a worrying and imminent requirement to pay its share of the development costs of the Ivanhoe and Rob Roy oilfields in the North Sea.

With production still a couple of years away Ewan Brown and I needed to find a way to reduce Pict's overall expenditure. The solution seemed to be to swap a number of Pict's North Sea blocks (on which commitments had been made to the Government to drill various expensive exploration wells) for a number of blocks onshore where drilling costs are considerably cheaper. After several months of negotiations we agreed a deal with the French company, Elf, and in the spring of 1987 that agreement was concluded. One of its attractions to Pict was that it brought an interest in areas of North Yorkshire where several small gas-fields had been discovered which in due course should be profitable.

The Elf cross-assignment would bring about some useful economies, but it was not the complete solution. Among the ideas which Ewan and I discussed was the possibility of a corporate deal, perhaps a merger with another company. One such company which swam into view was British Borneo, once an active company in the Far East and for many years now inactive, but with substantial shareholdings, particularly in BP and Shell. With its assured revenue and a full listing on the London Stock Exchange it seemed an ideal match. We met with British Borneo informally over dinner at the RAC Club in Pall Mall in August 1986 and spent a pleasant evening with Sir Douglas Morpeth, the Chairman, and his fellow directors. One of them happened to be Michael Rendle, formerly the head of BP in Australia during my years there. Attempted mergers generally fail, either because of the conflicting personal aspirations of the two sets of managers, or because they are unable to agree on the respective values of their companies. The Pict/British Borneo merger did not come about as we had hoped, but it was useful to know that here was a company wishing to awaken from its torpor, a matter I decided to follow up with Michael Rendle at a later date.

We considered other corporate deals to solve Pict's dilemma, but it was Amerada Hess which opened the way forward. Amerada had by now become the operator of the Ivanhoe and Rob Roy block, having acquired it from Monsanto. A deal agreed in 1987 between Amerada and Pict involved the former providing the five million pounds Pict needed to fund its share of the fields' development, in return for which Amerada would gain a stake of just below fifty percent in Pict. It was accepted that Liberty would continue to provide the technical management of Pict until 1989 when the two fields were due to come into production, at which time Pict would recruit and employ its own staff. From my point of view this was a very satisfactory outcome.

Chapter Sixteen

Song and Timothy left for Bangkok as soon as the Killearn Primary School term finished, and I joined them on Christmas Day. It was the middle of the cool season and the ideal time of year to be in Thailand. We stayed again with Nai and I was happy to enjoy again the simple pleasure of breakfasting on the veranda, with fresh papaya and a squeeze of lime, black coffee and fried eggs, feeling the growing warmth of the sun seeping into my winter-white body. A few days later, on New Year's Day 1987, we took another of our impromptu trips, this time to Burma.

Once a country rich from its exports of rice, oil and gemstones, two decades of 'Socialism the Burmese Way' had reduced it to poverty. We found a room at the Strand Hotel on the waterfront, by the brown, muddy Irrawaddy, bustling with sampans powered by Honda outboards. It has since been refurbished, but in 1987 our room was grubby, with a creaking ceiling fan, and its single twenty-watt light bulb made it impossible to read. Water leaked from the cistern of our shared bathroom to form a pool which spread under the door into the corridor. The rest of our expedition was in the same vein. A fourteen-hour overnight train journey took us to Mandalay where, at the last minute, we were spared having to stay in the most sordid and dirty guest house I have ever seen, when a room was eventually found for us at the Mandalay Hotel.

The journey by riverboat to the pagoda-dotted plain of ancient Pagan was, for me, the most memorable and enjoyable part of the trip. Wakened at the hotel at 4.00 a.m. we climbed into an ancient bus which rattled around the boarding houses of Mandalay picking up passengers, almost all of them back-packers from Australia, Israel, Germany and Britain. Scrambling with our luggage over the levée, we were turned away from what at first we assumed was our boat. We trudged along the turd-smelling sandy shore and boarded a two-deck riverboat reminiscent of pictures I'd seen of the Clyde-built boats which plied these waters in the pre-war days of the Irrawaddy Flotilla Company. There were deckchairs on the upper deck and we huddled down out of the chill air in as many layers of clothes as we could muster. At 5.30 a.m. we slipped our mooring and drifted into mid-river where the current caught us and bore us downstream at a good speed.

> *"Winding around pole markers planted in the river bed, through occasional swirling whirlpool areas, the water murky, but not dense as in Rangoon. Passing long sandbanks, some 20 feet high, new and unvegetated, some older with rice and fields of yellow rape. There were rafts of bamboo steered by three oarsmen at the stern, and huge rafts of teak logs with two or three thatched huts on them, pulled lazily downstream by a tug; and scattered sampans, upcurved at bow and stern, with fishermen casting their nets. And lines of fishermen pulling in their nets on the shore.*
>
> *Long stretches of the river bank are treeless, until we pass a village where palms and tall shade trees overhang the water. There are oxen hauling carts, and water buffalo (but with horns more tightly curved than those in Thailand), and occasional glimpses of distant hills beyond the plain.*
>
> *At seven o'clock, breakfast was served in the dining area on the lower deck, a meal of fried eggs, toast and jam, with tea or coffee. But it was impossible to escape the cold breeze, and not till nine was the sun warm enough for us to shed the extra shirts we'd put on.*
>
> *There were occasional village stops to drop or pick up passengers; more sampans, some with a simple sail and a cargo of melons. A lunch of excellent tomato soup and a curried meat and vegetable stew was served at 11.00 a.m., and later, high tea – a relic of past colonial times?"*

We arrived at Ngao-oo twelve hours later, where the villagers swarmed around us and, half in affection and half in curiosity, pawed and stroked at Timothy, much to his consternation. Thinking we would get to Pagan ahead of the back-packers, we climbed into a waiting trishaw and set off – only to be overtaken ten minutes later by a "Tourist Burma" bus full of our fellow passengers.

The three of us got sick in Pagan, which marred our enjoyment of the many temples and pagodas, but by a couple of days later we felt better. With my rucksack weighed down with chunks of silicified fossil wood we had found on the Irrawaddy beach, we piled into a crowded pickup at dawn for the four-hour drive east to the railway station at Thazi. Unable to get Upper Class seats we travelled to Rangoon in a carriage where, in the daylight you could watch the wheels through holes in the wooden floor, and after dark you kept your feet off the floor to avoid the scuttling cockroaches. I remember Song's comment on the bucking and lurching ride: "No wonder that in the early days of the railways they talked of riding the Iron Horse!" Not many women would have put up with the dirt and discomfort of that Burmese trip as stoically as Song.

She needed that stoicism. In the early spring of 1987 we detected a lump in Song's breast and our GP in Killearn referred her for a biopsy. I was in the USA, visiting contacts in Denver and California and our oil and gas properties in Oklahoma and Louisana, when I called her and got the dread news that she had breast cancer. A mastectomy was recommended and I accompanied Song to the Western General Hospital in Glasgow where she was to have the operation. With its confusing Victorian corridors and cluttered wards, its windows grey with grime, it is little wonder there were tears in her eyes when it was time for me to leave.

Back home it took a while for her to recover her strength, but by early May I thought Song was strong enough to enjoy a short holiday in Venice. A month after her operation we flew to Milan from where I thought it would be fun if we were to take the 'Super Rapid' train to Venice. It was a ghastly journey. Due to depart at 3.20 p.m. we were still standing on the platform an hour later. When eventually it left, the train was diverted via Bologna and we arrived in Venice five hours later. Doing my best to comfort Song, who was by now tired and hungry, we took the *diretto* to St Marks Square and found the small hotel where I'd booked a room. An argument with the receptionist ensued when she showed us to a cramped room above an alley, instead of the front room I'd asked for. Exasperated by now, I told Song to wait there, and I turned on my heel and strode out, determined to find a more suitable place to stay. Within a hundred yards I came across what seemed to be a pleasant hotel, entered, and asked the receptionist if they had a comfortable room. Yes, she said, and showed me up the marble staircase to a large room opening off the first floor gallery. I informed her that I'd take it and returned to collect Song. It was only later that I found that the Danieli Hotel, formerly the 14th Century Palazzo Dandolo, is one of the finest hotels in Venice, and one of the costliest. From that moment on we enjoyed our stay in 'La Serenissima', but it had not been a good start.

For some while Song had harboured a wish to become more actively involved in art. She was already talented at sketching and had been taking daytime extra-mural lessons, but she wished to take up painting and, if possible, enter the Glasgow School of Art. Just one week before her operation she had submitted her portfolio, and when we returned home from Italy we were excited to find that she had been accepted on the four-year Painting and Drawing degree course.

The school was enjoying a high reputation in the art world, and the building itself is something of a shrine, since it is considered the finest example of the work of Scottish architect and artist Charles Rennie Mackintosh. Song enrolled in the Autumn Term and began her studies.

Although she appeared to make a full recovery from the operation, Song was no longer as strong as she had been, and she could quickly become tired. We agreed it would be a good idea to buy a small flat near to the Art School so that she could relax there whenever she felt the need. We settled on a ground-floor, red sandstone, Victorian tenement flat at 54 Scott Street, on Garnet Hill and just a stone's throw from the school. It cost a little over £20,000 and when I put the deeds of the property in Song's name she was overcome with joy. She loved the place and devoted most of her spare time to furnishing and decorating it to her liking. As well as its intended use as a place for Song to rest, we found it useful as a *pied à terre* when we were in the city for a concert, or to make available to friends.

Taking advantage of Millfaid's closeness to Loch Lomond I indulged my love of boats and bought a canoe, a dark green Canadian style craft with varnished mahogany trim – a handsome boat – and we enjoyed voyages down the Endrick to the loch, and trips to barbecue and camp on its many islands. But the most memorable was a trip that Timothy and I took down the River Tweed in the summer of 1987. Song drove us to Kelso where we manhandled the canoe off the roof of the Volvo and eased her into the water, before stowing our tent and other camping geart on board. Timothy was only seven but soon got the hang of paddling and jumping overboard to help drag the boat through shallows or lower her gently over weirs. We found a grassy bank on the English side to pitch our tent that night and after cooking our dinner we walked to the nearby village to telephone Song with our progress. It was great fun and when we reached Berwick the following afternoon we wished the Tweed were several times longer.

While we'd been camping on the river bank Song was spending the night at Sunlaws House, a comfortable Borders country-house hotel some miles distant. It was a useful reconnaissance, and later that year we chose the same hotel for the first of Liberty's Christmas parties, a residential weekend for staff and their families.

Since crewing Gordon's yacht down the West Coast of Scotland my ambition had been to own a yacht of my own. Something around a twenty-footer, I thought, would be suitable, and I set about studying the yachting press and advertisements in the Scottish papers. Then one day while Song and I were looking around Kip Marina on the Clyde coast we came across *Ashanti,* a solidly-built Bermudan-rigged sloop with a price tag that was within our reach. Down below she had a forward cabin and a saloon, a sensible-sized galley and a 'heads', and more varnished mahogany lockers and stowage cubby-holes than I could count. At 34 feet she was admittedly a little longer than I'd had in mind, but I decided she was the boat for us and we bought her.

I still have the log-book of the many short and long cruises we made in *Ashanti,* first based on the Clyde at Kip, and at other times based at Craobh Haven south of Oban. Our longest cruise that year, 1988, was with Ian and Julia Rolfe when we explored Mull and Skye before leaving her at Arisaig and taking the train south for a few days break. When the train reached Fort William I telephoned Marjorie, my assistant, and asked her to arrange for a taxi to meet us at Dumbarton Station to take the five of us back to Millfaid. We got there to find not a taxi but a stretched limousine, one of those ridiculous vehicles beloved by popstars, with a cocktail cabinet and TV antennae ostentatiously mounted on the back. Marjorie had been unable to find a normal taxi that was large enough, she explained with embarrassment. Sitting inside in our dirty, salt-stained clothes, with grimy hands and matted hair, we reclined with a gin and tonic each and laughed all the way home, waving like royalty to bemused bystanders as we went.

I learned on that cruise that the Rolfes are not keen sailors; in fact Ian's preference was to pass the time with his nose in a book from the moment we left one anchorage until we dropped anchor that evening in another. I learned too that *Ashanti,* while a fine sea boat, had a number of troublesome mechanical defects. One such problem was the control linkage to the gearbox, a problem which nearly caused an expensive accident one day when she failed to engage neutral gear while manoeuvering in a crowded lock on the Crinan Canal.

That winter I had a boatyard on the Clyde strip her down and carry out repairs, but her mechanical problems were never fully overcome. We sailed again on the West Coast the next summer, with a variety of minor difficulties caused by faults in the mechanical and electrical systems. But the most serious mishap with *Ashanti* occurred the following year and was caused not by mechanical problems but by my own incompetence. In June 1990 we were nearing the end of a cruise which took in Ardnamurchan and Plockton and included a call on old friend Sir Iain

Noble at his home on Skye. I had persuaded Ian Rolfe to give sailing another chance, and the other members of the crew were my brother, John, and an architect friend named Peter who would eschew our nightly dram and curl up in his bunk to read his Bible – on reflection, perhaps a sensible precaution bearing in mind the Skipper's ineptitude. There were magical moments as we explored anchorages around Mull, particularly Tinkers' Hole where granite cliffs encircle a crystal-clear pool of calm water and you can hear the distant roar of the waves crashing on the rocks which guard the anchorage's treacherous entrance.

Disaster struck the next day after running for several hours before a westerly wind which took us the length of Mull toward the mainland. The wind was behind us and conditions were perfect for putting up the cruising chute, a lightweight and colourful sail which billows over the bow like a parachute. We sailed this way in grand style until the time came to change course, as we intended to make a landfall on the Garvellachs, a group of small islands of reported archaeological interest lying southeast of Mull. I made the elementary mistake of starting the engine and turning the boat under power until she was heading into the wind – a reasonable manoeuvre if we had been intending to drop the mainsail or the foresail, but, as I quickly learned, a silly thing to do before dropping the cruising chute. With me at the helm I gave the order to drop the cruising chute, and immediately the lower part of the sail and the attached lines dropped into the sea and were carried aft. There was a thump, the engine stalled, and I found I could not turn the helm. I tried starting the engine again but the problem only got worse, as it wound the sail and ropes more tightly than ever around the propeller. My first thought was that we were now being carried by the tide toward the infamous Gulf of Corryvreckan, the narrow seaway between the isles of Jura and Scarba where, in certain weather conditions, the whirlpools and standing waves can easily swamp a yacht. And if the Gulf didn't sink us, then the submerged rocks and hazards along this coast certainly would unless we could regain control soon. Our problem was compounded by the mass of trailing sail which filled like a drogue and put us totally at the mercy of the water currents. Obviously we needed to clear the propeller and the rudder as best we could. With someone holding my ankles I hung over the stern and hacked away as much of the debris as I could with a sharp kitchen knife. At last I cleared the rudder and got rid of the ballooning mass of sail. I couldn't reach the prop and so the engine was still completely frozen, but at least we could resume sailing.

If there is one stretch of water on the West Coast where it is advisable to use the boat's engine it is in the powerful tidal streams and rock-strewn waters of the Sound of Luing. And to get back to our berth at the marina of Craobh Haven we needed to travel the length of the sound, against the tide. After rounding the Pladda lighthouse which marks the north end of the sound we turned southwest to meet head-on the north-flowing tide. It took a couple of hours to inch our way through the eddying currents of the sound and there were times when, although moving fast through the water, we seemed to be making no progress past the islands on each side. At last we reached the southern end of the Isle of Luing and entered quieter waters. The entrance to the marina is narrow and normally a skipper would not dream of entering and making his way to his berth under sail alone but we managed, and with feelings of relief we came to a stop at precisely the right point on the pontoon. It was my job, of course, to strip off and plunge into the sea with the knife to clear the last remaining bits of sail and rope from around the propeller shaft. It took twenty minutes before the last tightly-packed strands were off, and I emerged with a seriously lowered body temperature and an uncontrollable shiver. Wrapped in sleeping bags while the crew plied me with hot drinks and massaged my limbs, it was a further half-hour before my shivering ceased and I felt my core temperature was back to normal.

By 1989 Pict's North Sea oilfields, Ivanhoe and Rob Roy, were in production and the agreement whereby Amerada Hess had funded their development came into effect. Pict was now to employ a staff of its own and so Liberty's management agreement was concluded. It no longer made sense for Croft to employ Liberty's services and therefore I sold the business to Croft. Liberty's staff, including myself, became employees of Croft, and Liberty itself remained simply my vehicle for holding shares in Croft.

In truth, even if Pict had chosen not to go its own way, the time had come for Croft to become a separately managed company. It had benefited from its junior-brother relationship with Pict in the early days, but it had grown and raised its profile to the extent that the arrangement was no longer tenable. A fellow-director of Iain Harrison's in his shipping company was Peter Wordie, son of the geologist James Wordie who accompanied Ernest Shackleton on the ill-fated Endurance expedition to the Antarctic. For Peter, Croft struck a chord and he procured an investment of a half-million pounds by his family business, which raised the company's capital to one million. Shortly after that, as a preliminary to seeking further investors, we had restructured the company and changed its name to Croft Oil and Gas plc. We had expanded our onshore UK portfolio to six blocks – in Northern Ireland, Scotland, the Pennines and eastern England – and by judicious farming out we had participated in boreholes on two of them. We had carried out our first seismic survey as operator in our eastern England blocks, and offshore we had been awarded two blocks in the Southern North Sea gas basin. We had also taken our first step to becoming an offshore operator by applying successfully for block 48/4a in the same basin. (I had stayed in touch with Michael Rendle of British Borneo, and in this recent licence award I took some satisfaction from having persuaded him and his formerly sleepy company to join us – its first foray into the exploration business).

Overseas we were expanding too. In 1988 we took advantage of my earlier experience in New Zealand with BP and applied successfully for a large onshore exploration block in the Hawkes Bay district of the North Island. We then introduced partners into the venture and began seismic work which revealed a major prospect.

Toward the end of that year I happened to read in the Financial Times that the government in Burma was opening its economy to foreign investors and that it planned to invite oil companies to apply for exploration blocks in the country. Until the early 'sixties Burma Oil Company was a significant oil producer there, but following nationalisation there had been no foreign investment and the perception among the world's oilmen was that major oilfields remained to be found. Quickly mugging up on the geology of the country I sent off an expression of interest to the Burmese Government in Rangoon, copying it to the Commercial Attachée at the British Embassy there. A month went by, Christmas came and went, and still I received no reply. I contacted our embassy again and asked them to investigate the matter. Within days we received a telex from the government asking me to meet them for discussions the following week. This was impractical for several reasons, one being a planned family trip to explore Andalucia, but we agreed that I should make a visit to Rangoon in the spring.

In the office we busied ourselves preparing a glossy presentation, compensating for our lack of detailed data on the petroleum geology by including impressive maps of Southeast Asia showing its plate-tectonic elements, and emphasising the similarity between the Irrawaddy Basin of Burma and the oil-productive onshore basins of Indonesia. The day after I returned with the family from Malaga I left for the Far East. Armed with the technical presentation material we had prepared, and with several bottles of whisky with which to bribe officials, I flew via Bangkok and arrived in Rangoon on the 10th April.

In the two years since my last visit, the Strand had been restored to something resembling a comfortable hotel, although the touts still loitered in the shadows outside, offering to buy your US dollars for several times the official rate. I called at the Embassy and met Martin

Moreland, the Ambassador, and was pleased to find him and his staff interested in our venture and very willing to assist. Perhaps it helped that Martin had spent a couple of years in the City, seconded by the Foreign Office to work with my old friends Clive Hardcastle and David Robertson. Martin and his Commercial man, Fraser Wilson, explained the delicate relationship between the UK and Burmese governments, where the UK was officially ostracising the Burmese in protest at the recent heavy-handed treatment of pro-democracy activists, while on the other hand maintaining close informal links with influential officials. I learned too that I was by no means the first oilman to call on the Burmese Government: BP had been there recently as well as British Gas, BHP, a French company, several Japanese and, of course, a number of American oil companies. So much for my thinking I was stealing a march on my competitors!

The following day I took a taxi to the offices of the Energy Planning Department and introduced myself to the committee of some ten officials who would hear my spiel. The Director General and chairman was U Tin Tun, a cultivated and courteous man who had worked his way up the ranks of the civil service. Others gathered around the long table included an attractive middle-aged woman lawyer, a couple of government geologists (one, a recent graduate from Aberdeen), a charming avuncular pipe-smoker, and several uniformed military men. The atmosphere was relaxed and cordial in a way which surprised me, since in neighbouring Thailand such a meeting would have been stiff and formal. Perhaps it is not too much to suppose that it came from a long history of working with the British, albeit not recently.

I had rehearsed my presentation, of course, and began by suggesting that although Croft is a small company its very smallness might be seen as a virtue. After all, I explained, if a major oil company discovers a modest-sized oilfield it may not give it a high priority for developing and exploiting, whereas if Croft were to discover the same oilfield it could transform our company and we would certainly give it a very high priority indeed. Tin Tun's response encouraged me: "Ah yes, Dr Ridd" he said, "how right you are. As that fellow-countryman of yours, Schumacher, has pointed out, small is beautiful!"

A few weeks later we heard that our approach had been successful and we were to be offered a large block of acreage centred on the small town of Henzada at the head of the Irrawaddy delta. We were astonished and delighted to have been selected along with only eight other oil companies, including household names like Shell, Amoco and BHP. My flights to Rangoon became a frequent occurrence as we negotiated the details of the production-sharing contract and the work-programme, and on my visits I was invited by the British Embassy to stay at their guest house. This was significantly more comfortable than the Strand Hotel, as it offered pleasures such as a swimming pool and a canteen which served those delicacies missed by the British abroad: beans on toast, and sausages and chips.

Our small full-time staff in Glasgow could no longer cope with this surge of activity and I was fortunate to be able to call on friends from my Britoil days who were now consultants, including the very personable geologist, George Farrow, and the silver-tongued and urbane commercial expert, David Montagu-Smith. I enjoyed those visits, largely because of my gracious treatment from the Burmese and their straightforward business conduct. Yes, it was a dictatorship with a reputation for treating its citizens harshly, but neither in Rangoon nor up-country did I see any evidence of it. There was certainly poverty on all sides, and once in Rangoon I had the chilling sight of a column of half-naked prisoners, chained together and with leg-irons clanking, being marched down a side street. But I never saw any sign of corruption and felt no guilt for my dealings with this pariah state.

While progressing matters with Rangoon, it was also urgent that I find a larger oil company to farm-in to the venture and join Croft in accepting the award. As I have said, the award of acreage in Burma was viewed as a considerable prize, and there was no shortage of companies willing to visit our office in Glasgow to discuss the possibility of joining us. The company we agreed to proceed with was Clyde Petroleum. Clyde had been established in the 'seventies by a Glasgow merchant bank (much as Pict had been) and was run by Colin Phipps, a fellow UCL geology graduate with whom I had overlapped in the mid-'fifties. His Exploration Director was another UCL alumnus, John Martin, with whom I had worked in BP over many years. (John had once been a close friend, particularly when we had worked together in Libya and the USA. We continued to have frequent contact through the rest of our respective careers, but for reasons I shall not go into I could never feel the closeness to John that we had once enjoyed.)

Under the agreement with Clyde, we set up a new company called Croft Myanmar Limited in which Clyde held 90 percent of the shares and we held 10 percent. Clyde agreed to reimburse all of Croft's past costs and to pay our ten percent of all future costs including seismic surveys and exploration drilling. It was a very satisfactory outcome from what had started with the casual reading of a newspaper article. Of course, the big test of the venture was whether we would discover oil or gas within our block. Clyde's performance as an operator was less than impressive, but seismic surveys gradually revealed that a very large prospect was present. In Clyde's and our opinion this certainly justified pursuing, and plans went ahead to import a rig and drill an exploration hole near a village called Lemyethna. As the rig was being mobilized I took the opportunity to visit the drilling site, flying first by helicopter from Rangoon Airport to the town of Henzada on the west bank of the Irrawaddy. To my surprise the authorities raised no objection to our route, which took us low across a large star-shaped building on the edge of Rangoon, the notorious Insein Prison.

The prospect which our seismic surveys had defined was so large that it had the potential to transform the Burmese economy and to make Croft's shareholders very wealthy. As the drilling bit dug deeper the daily telexes would arrive at our Glasgow office, containing the latest information on the well's depth, the strata it had penetrated, and any indications of oil or gas. But slowly our hopes gave way to disappointment. The well reached its planned total depth and we had to accept that it would not make the oil discovery for which we had all held our breath. Our only consolation was that none of the other companies awarded onshore blocks at the same time as Croft found any commercial quantity of oil or gas either. The oil industry had assumed that in the quarter-century since Burmah Oil had been forced to leave the country, the Burmese Government had been sitting on its hands; but the truth is they had done a workmanlike job of exploring their country and had found anything worth finding long before the arrival of Croft, Shell, Amoco and the others. The well's failure was harder on Clyde than on Croft, as they had borne the entire cost of the venture, a blow to their balance sheet which caused many a raised eyebrow in the City.

Throughout the excitement of the Burma venture, life at Millfaid had continued happily and harmoniously. We relished the passing of the seasons: there were snowfalls which at times would block our way to Glasgow; the Catter Burn would occasionally overflow and inundate our front field; and on 13th February 1989 a ferocious wind which became known as the Glasgow Hurricane flattened trees in the forest behind the house, but we escaped without damage. They added interest to our already busy lives. Song remained absorbed by her studies at the School of Art, but we agreed she needed a place at home in which to paint, and so we had a large studio built adjoining the garage block. Again, we asked John Boys to design it and oversee its building, and when finished it sat as unobtrusively and comfortably in the landscape as Millfaid itself, with its Highland slate roof and walls of white harling and red sandstone.

That March I was confined to bed for over a week with severe backache, the first time in my life I had experienced a pain so disabling. More accurately, I was confined to the floor, as it was only when lying flat on the floor with my legs up on the bed that I could relieve the pain in the small of my back and my right leg. But that problem was forgotten in the middle of 1989 when we were told by Song's consultant that she had a recurrence of breast cancer. For two years she had been fit and apparently healthy and had coped bravely and confidently with her mastectomy. In June she was readmitted to the Western General Hospital – that disgracefully decrepit place – for a second operation. Tests showed that the tumour had not been eradicated and in October she had yet another operation, followed by radiotherapy. It was a ghastly setback, but gradually Song recovered from the disappointment and regained her strength and her indomitable spirit. She resumed her studies and joined the Glasgow Art Club where she became a well-known and liked member. Under its somewhat shabby exterior, Glasgow has a rich cultural life which we grew increasingly to enjoy. The new Concert Hall attracted world-class performers and, on a smaller scale, the Pollock Art Society held monthly recitals in the intimate surroundings of the library of Pollock House, followed by friendly black-tie dinners in the candle-lit Victorian basement kitchen.

Early in Song's illness, when her spirits and her strength were low, her sister Nai agreed that Pee Tau, Nai's housekeeper in Bangkok, should come and live with us at Millfaid to act as *au pair.* Pee Tau herself was delighted at the prospect of seeing a new part of the world, and the Home Office put up no objections and granted her a temporary visa. It proved wonderfully liberating for Song, particularly as they could converse together in Thai, and Timothy – who was by now completely bilingual – took to this new family member immediately. Her grasp of English was rudimentary at first, but she picked up the language quickly and her natural warmth made her a favourite among our friends. Over the years that followed, Pee Tau was to play an important part in the life of the family.

Perhaps because of the time he had spent in Thailand throughout his young childhood, Timothy was a proficient swimmer. This was something that Song recognized and decided to develop, and throughout the late 'eighties Song would drive hundreds of miles a month taking Timothy to his club in Milngavie and to swimming competitions in the Glasgow region. We also began to investigate a suitable school for Timothy to attend when he was a few years older; in the meantime he was a pupil at the preparatory department for the High School of Glasgow and due to move up to 'Transitus' when he became ten. We visited most of the better-known boarding schools in Scotland – Glenalmond, Strathallan, Gordonstoun – but it was Loretto, in

Musselburgh, that we settled on, attracted mainly by its impressive headmaster at that time, Norman Drummond. We duly put down Timothy's name, to begin there when he reached the age of thirteen.

Soon after Christmas 1990 the three of us flew to the USA for a winter holiday in the sunshine. It was cold and snowy in Boston where we broke our journey, but in Florida, where we planned to spend the next fortnight, the weather was warm and sunny. We hired a gas-guzzling Buick at Orlando Airport and checked into a motel on International Drive, then spent the next week enjoying the slick delights of Disney World, Sea World and 'Wet and Wild,' and sampling the scores of 'Kids Eat Free' restaurants along with thousands of other tourists. We drove east to the Kennedy Space Centre, and continued down the coast through Palm Beach to Miami and the Everglades. We hired a motorboat there and explored the reed-beds and the narrow waterways, watching ospreys flying from their tree-top perches, and searching for the trademark creature of the Everglades, the alligator; we came across several but the biggest one we found was a mere four-footer. It was an excellent holiday, carried out, as usual, *ad hoc,* and Song remained in good spirits and energetic throughout.

She did not mention it at the time, but back at Millfaid Song revealed that on the flight to London she had felt her neck and discovered pea-sized lumps beneath the skin. It was worrying and we arranged to meet with her consultants, Adrian Harnett and Professor George. Their opinions were not reassuring. They were virtually certain that the lumps were secondary tumours, but that neither surgery nor radiotherapy would affect the long-term course of events. It was agreed that she should have full-body scans, and that if they showed further tumours then the appropriate treatment would probably be the drug tomoxifen. As we returned to our parked car it was as much as we could do to stifle our tears of disappointment. It was as if some almighty power was playing cat and mouse with us, cruelly grasping us in its claws only to release us, before catching us again, repeatedly.

It was late March before the results of the scans and other tests were available. Professor George explained that they showed shadows on both lungs which could be tumours. He gave Song the choice of tomoxifen or chemotherapy, explaining the pros and cons of each. To my surprise Song said that she wished to accept neither treatment for the time being, and it was agreed that she should return in two months for a further check.

Perhaps to demonstrate her disdain for her disease Song said she would like a short holiday in the Highlands. And so the following week the three of us drove to Skye, to revisit Skeabost House, the fishermans' hotel that we had enjoyed a dozen years earlier. From there we moved on to Mark and Geraldine Irvine's excellent Summer Isles Hotel. That is golden eagle country and our constant scanning of the skies was rewarded with the sight of a pair of these magnificent birds.

Through all of her travails Song continued her painting, but a strange anger started to appear in her work. Self-portraits from that time show her with a severe expression, while others portray her looking at her own reflection in a mirror where the image is of herself, but in the form of a sad clown. I can only guess at the state of mind which prompted these unsettling paintings.

At the end of May we spoke again with Song's consultant. Recent x-rays showed that things were a little worse: there were swellings in the lymph glands around the branchings of the tracheae, as well as some fluid on the right lung. He recommended again that Song should take tomoxifen, saying that it can keep things static or bring improvement for a year or two. She refused. Instead, she said that she would like to return to Bangkok for a holiday.

We flew to Bangkok on 22nd June. As a surprise I had booked three Business Class tickets for the family, hoping to see the pleasure on Song's face as I walked them into BA's Executive Club lounge at Heathrow. But instead she reprimanded me, saying that this surprise had deprived her of the pleasure of anticipating the trip. Yes, I thought, Song has lost none of her old spirit!

I stayed for a week with them in Bangkok, taking the opportunity to make a number of business calls on other oil companies and on the Department of Mineral Resources with whom we had been in negotiation for some time over the award of a number of exploration blocks in Thailand. Song and Timothy remained, enjoying being back in the bosom of her family and particularly that of her sisters, Nai and Ying.

We spoke frequently on the telephone when I was home again at Millfaid. All seemed to be well until, about a fortnight after I had left her, she complained that she had become paralysed in her legs. I rushed back to Bangkok immediately. I can still see her in my mind's eye, sitting on the balcony of Ying's house where someone had carried her. She greeted me with tears in her eyes, tears of sheer disappointment that things were going so badly wrong. Now, fifteen years later as I sit alone typing this, tears again well up in my own eyes. She was also in distress over her inability to pass water, although that was quickly dealt with at the nearby Samitivej Hospital, a model of caring and efficiency. The next day, Sunday 14th July, I took her to Bangkok Airport and carried her in my arms onto the BA jet that was to fly us back to Britain. A wheelchair was waiting for us at Heathrow for the connection with the flight to Scotland, and when I carried her off the plane at Glasgow Airport an ambulance was on the tarmac to take her to hospital.

Radiation treatment of the neck and upper back commenced immediately, since her myalogram indicated nodular growths on the lining of the spinal canal. There was no improvement, and a week later she was advised to begin weekly intra-thecal chemotherapy; while unpleasant in its side-effects, she was advised, her health would deteriorate gradually if she failed to have the treatment. Song was not inclined to undertake the treatment and told her consultants so. To our surprise they responded that they thought this was the right decision, but whether they said that to comfort us we didn't know. The fact is, we were comforted.

The decision made, Song seemed relieved and in good spirits, saying she was happy and excited over my future as she thought of my business prospects and of my re-marrying. She was pleased too that her sister, Tari, was arriving soon from the USA to be with her. It was arranged that an ambulance would bring Song to Millfaid the following day, the 25th July. Tucked up in bed she asked me to bring her favourite pictures from around the house and re-hang them in the bedroom, pictures by Jack Knox, Anne Gordon, Anne Anderson and Emily Bakker. Our local doctor, Stewart Cumming, attended her with great care and gentleness and a kind Macmillan Nurse visited.

On Song's last afternoon she complained of a tightness in her chest and difficulty in breathing, and she was frightened she may suffocate. As Dr Cumming gave her a further shot of diamorphine hydrochloride she asked for her sister, Tari, and said "I want to say goodbye." As she drifted into sleep she answered me that she felt no pain and was content.

Song died at 3.20 a.m on the 2nd August. My little wife of seventeen years had gone.

Chapter Seventeen

Nothing was the same again, of course.

We were fortunate that we had Pee Tau with us to continue looking after our domestic arrangements. She was devoted to Timothy and his loss would have been even harder to take if she had not been there. Timothy by this time was a pupil at the High School of Glasgow and I had to engage a driver to drop him there and pick him up at the end of the school day. It is an excellent school but we had planned a boarding-school education, and Timothy started at Loretto in 1993.

I thought he had settled well in Hope House until one morning in the Winter Term I received a call from him. I was in bed still at the time. "Where are you calling from?" I asked. He explained he was in Glasgow with a school friend, the two of them having stolen out of their dormitory at five o'clock that morning. "OK" I said, "meet me at my office at nine o'clock and we'll talk about it." We met and the boys listed their grievances, none of them very serious, and I undertook to write to the Headmaster about them. I told them they could spend the morning in Glasgow and that we would meet for lunch at a favourite pizza restaurant, after which I would put them on a train back to school. Having got his grievances off his chest Timothy happily accepted this ruling and went on to enjoy his school career, telling me years later that he would not have missed it for anything. He meant, of course, the social life at Loretto, not any academic excellence it may have had.

I had come to terms with being a widower and attending social gatherings on my own. In fact since Song's death it was noticeable how few invitations I received, bringing home to me just how much I had relied on her for maintaining our active social life. But in the spring of 1993, nearly two years after Song's death, I was called by Sandy Stewart. I had got to know Sandy through his being the Honarary Consul for Thailand in Scotland. He explained that the Ambassador would be visiting Glasgow shortly, and he wondered if I would be willing to join a small dinner party that he would host at the Western Club. The seven guests, including spouses, were drawn from industry and academia and it was a pleasant occasion. I found myself seated next to the Embassy's Second Secretary, Khun Atchara, whose job it had been to organize the Ambassador's Scottish trip. When she had overcome her surprise at finding I could speak some Thai we had an enjoyable evening, laughing and talking together. A few days later I contacted her at the Embassy and asked if she would care to meet for dinner when I was next in London. Not yet forty, Jair (for that was her nickname) was tall and elegant and we shared a number of interests. She agreed to meet again and our friendship blossomed into a close romantic relationship which lasted for the duration of her London posting, and for a further year or two after that.

Initially, and predictably, Timothy was unhappy that I should have found a replacement for his mother, as he saw it. And even Pee Tau made things uncomfortable for me, but driven by different emotions. Now, whenever I told her that I was going to London on a business trip, she would grin knowingly or make inappropriate comments about my true reasons for going south. Perhaps without realising it I felt guilty to be in love again, which may have made me over-sensitive to her silly comments. At all events I didn't see why I should have to put up with them. For the several years that Pee Tau had been with us I had written annually to the immigration authorities to seek extensions to her visa. But in light of what seemed her growing possessiveness I decided that the time had come for the family to make do without her. She returned to Thailand where, with her grasp of English and her culinary skills, she had no difficulty in finding a good job.

Just before I met Jair, Timothy and I made a couple of journeys to the South Coast to examine yachts for sale. I had sold Ashanti several years earlier, tired of her apparently insoluble mechanical and electrical problems. Now I was pining for a replacement. I settled on one of Northshore's range of yachts, the Vancouver 28, a sturdy cutter-rigged vessel capable of sleeping four persons. We supervised her building at Northshore's Itchenor yard and Timothy and I drove from Millfaid to take delivery of her in April 1993. We named her *Tethys of Lorne*. With her tan sails, fresh teak trim and black and white hull, she was as pretty as a picture, and after a weekend getting used to the feel of her around

Chichester Harbour we reluctantly tied her up at Northshore's jetty until the end of the month when I would return with friends to sail her to the Clyde.

That turned out to be an eleven-day test of her – and our – toughness, often battling into the teeth of Force 7 headwinds, a test which she passed with flying colours. John and Paul Truitt were my crew as far as Falmouth and on our first day out we found ourselves in dense fog. Each of us was as nervous as the other as we felt our way through the Solent, with the boom of fog-horns telling us of invisible ships passing on all sides. By the Needles the fog had cleared and we sailed on. Then in Lyme Bay an exhausted racing pigeon came aboard, apparently blown off course by the strong northerlies. It huddled with us in the cockpit through all the following night till we made our landfall, when it flew up to a rooftop. A ring around its leg said "I am lost, please call Ashington ...." which we did.

Colin and Gordon joined me for the next leg, and after some poor navigation on my part which nearly caused us to run onto the Manacles reef we rounded Lands End and sailed north into the night. Dolphins played around the bow in bright moonlight as, well out of sight of land, we headed for the western tip of Wales. It was a bright and warm spring day and children were diving from the harbour wall as we anchored in the approach to old Fishguard. A couple of beers at the waterside Ship Inn followed by fish and chips up the hill in the new town rounded off the day.

Paul left us here by bus for Haverfordwest and home to London, and the three of us continued north, Colin, Gordon and myself. It was now seven days since leaving Itchenor, and we sailed on, close-hauled and assisted by the diesel engine. The northerly wind strengthened the next day as we approached Holyhead and, combined with a north-going tide, we found ourselves in rough water with ten-feet deep, steep-sided hollows into which our small boat would drop with a resounding crash. Conditions eased considerably as we entered Holyhead harbour. We found a vacant visitors' buoy and rowed ashore to shower and dine at the friendly and welcoming Holyhead Sailing Club. Gavin and Alastair joined us here, replacing Colin. Next day at dawn we pushed on north. The Irish Sea was rough and none of us felt completely well, but Alastair must have eaten something which had disagreed with him the previous night and spent the day in his bunk, a bucket tethered nearby. We were glad to reach the shelter of Port St Mary on the Isle of Man where we tied up alongside a fishing boat and went ashore in search of refreshments. But it was a town devoid of life on a Sunday night and we returned to *Tethys* and had an early night.

Another day of near-gale and gale force winds with rough seas brought us up the west coast of the Isle of Man and into the welcome haven of Portpatrick. We had got as far as Scotland and we celebrated ashore, wining and dining in the lively Crown Hotel. The next morning we were forced by the NE gale to tack our way along the Scottish coast, but as the wind veered we could sail on a broad reach, and as darkness fell and the stars appeared we could see Ailsa Craig slip by, and Arran, and at last we were tying up at Kip Marina.

Sailing had to be fitted around my continuing work to develop Croft, although I found time also to pursue my professional interests. I was appointed an external examiner for the postgraduate course in Basin Dynamics run by Royal Holloway College, and my association with the British Geological Survey continued. I recall also an interesting discussion meeting I chaired at the Geological Society in which Professor Sammy Gold defended his theory that oil and gas are derived from deep crustal processes and not from the effects of heat, pressure and time on organic matter within sedimentary rocks as all the world's petroleum geologists accept. And the Royal Society of Edinburgh elected me a Fellow, although I participated in few of their activities and in later years I let my membership lapse

But I must explain, even at some length, what became of Croft. Our list of shareholders had grown considerably by now. A private-placing of shares in 1990 brought in the investors Norwich Union and Murray Ventures. Two years later Croft agreed to purchase from Pict its small interest in the Claymore Oilfield and raised finance from the 3i Group, Hambros and from Pict itself. The thinking behind that acquisition was that it would provide us with a steady income which would fund our further growth. Although since the disastrous 1986 collapse of the oil price we had learned to live with prices below $20 per barrel, we were unprepared for the surprise sprung on us by Occidental, the operator of the Claymore Field. It was Occicental, of course, which had been operator of the Piper Alpha oil platform which had catastrophically caught fire in 1988 with the loss of 167 lives. With pressure on them now to make good their tarnished image they decided to build a new accommodation platform alongside the existing Claymore production platform to provide a safe refuge for personnel in the unlikely event of a fire. We and other partners argued against this extravagance, proposing instead less costly options, but we were outvoted. The new platform was built and the cost implications for the Claymore partners, including Croft, were enormous and over the short term destroyed our hoped-for income stream.

But there was better news on the exploration side of our business. As a result of our enhanced profile within the industry from our earlier North Sea licence awards, and I daresay by our successful project in Burma – notwithstanding our failure to find oil there – we were invited by two major North Sea operators, Conoco and Mobil, to join their groups applying for acreage in the Twelfth Round of offshore licencing. In 1991 the government announced the awards and our groups were successful in obtaining no fewer than eight blocks or part-blocks. Seismic work was followed by drilling, which confirmed that our acreage held extensions of BP's Donan Oilfield and Arco's Blenheim Oilfield. An exploration well in the same group of blocks discovered an independent small oil accumulation, 15/20b-11. Further south in the North Sea our operator Mobil drilled an exploration well on a hitherto undrilled salt-dome prospect where we had more success, discovering the Kyle Oilfield.

At about the time that we were enjoying these successes in the North Sea, the Liverpool Limited Partnership, a US venture capital fund, purchased part of Murray Venture's shareholding and injected further funds through the underwriting of a rights issue. And around the same time Harrisons underwent a financial restructuring and shares in Croft were transferred to Harrisons' shareholders, which brought onto the Croft share register a several well-known names. Although I was able to participate to some extent in these rounds of fundraising, my financial stake in Croft inevitably declined – whether through my direct shareholding or through Liberty – and by the mid-nineties it was not much more than seven percent. But importantly Croft now had a substantial share register which included a number of major financial institutions which were household names.

Overseas we were encouraged by our successful application for acreage in Burma, and I decided to apply for acreage in the adjacent country, Thailand. Having worked there in the 'sixties and become acquainted with the geology, the regulatory and tax regime, as well as several officials in the Department of Mineral Resources, it seemed a logical next step in our expansion. We were well pleased with the outcome of our application when we were awarded three blocks. Two were onshore in Southern Thailand where there were geological grounds for thinking there may be an extension of the prolific Gulf of Thailand Basin, and the third was an offshore block south of the holiday island of Phuket and extending as far as the median line with Indonesia where a number of gas and oil discoveries had been made. We had successfully negotiated quite modest work obligations on these blocks, but even so we needed to find a way to fund the ventures – we certainly could not do so from our own resources. We knew that BP had a couple of small onshore oilfields (which they had discovered on acreage they acquired when they bought Britoil) to the north of Bangkok, and we suspected that they were too small to be of key importance to BP. And so we contacted their management in London and negotiated a price at which they would be willing to sell them to us. The revenue generated from these small fields, we calculated, would be more than enough to cover the cost of exploration on our new blocks, and the residue would fund future growth. The vehicle for this project was to be a new Thai-registered oil company. We needed to raise US$25 million, and in July 1992 we began the search for investors.

It was, we thought, an elegant package which deserved to succeed, but we failed to find investors. Or rather, we failed to find them in the time available. We had been awarded the blocks nearly two years previously and the DMR was becoming impatient. I think in fact we would have succeeded but for the head of the DMR, a man named Sivavong who had little time for small oil companies and their need for imaginative ways of funding their activities. One of the government's requirements was that companies should capitalize their Thai subsidiary with a sum sufficient to cover the exploration work they had undertaken to carry out – in effect, this amounted to depositing a bond. Sivavong was used to dealing with major oil companies and could not understand that for a small company to lock up several million dollars in the way he proposed was difficult if not impossible. We failed to persuade him of the folly of that policy and so had to abandon the project. It was disappointing for us, of course, but from the Thai perspective too it made little sense – those particular onshore and offshore blocks which we were willing to explore have remained unexplored to this day.

I have written this as if I alone were responsible for such success as Croft was enjoying, but that was certainly not the case. Although our staff remained very small – generally about seven – they included three key people. I have mentioned Jed Armstrong, our geophysicist. He had shown that, single-handed, he could organise, commission and supervise a seismic survey and then interpret the results, and it gave great credibility to our approach to different governments that we were able to demonstrate that we had these capabilities in-house. Iain Patrick, our Finance Director, was equally vital to our success. Although not yet thirty years old, he was the brain behind our commercial thinking and it was Iain who carried out all of our financial modelling and budgetary control. And then there was Marjorie Beattie who relieved me of the need to concern myself with all aspects of running the office and employing the staff.

With our well in Burma drilled, and the collapse of our venture in

Thailand, we began to cast around for another part of the world where we could take advantage of what we persuaded ourselves to be a certain ability to obtain exploration licences from governments. We settled on the Central Asian republic of Kazakhstan. It had become independent in 1991 and was well-known to have been one of the Soviet Union's principal oil and gas regions. Again, we set about researching every aspect of its upstream business: the petroleum geology of the various basins, the licencing and oil-production regulations, taxation, export routes, legal system, and so on. When I felt that we were ready I wrote to the British Embassy and to the Kazakhstan Ministry of Geology in Almaty. The replies were encouraging and so I set off, again bearing gifts of Croft-engraved penknives and bottles of Scotch whisky. It was March 1994 and piles of heaped-up snow made it difficult to walk around Almaty. But it was not an unattractive city, with broad boulevards lined with tall poplar trees, and filling the southern horizon the towering snow-covered Tien Shan Mountains. I checked into the Dostyk Hotel, a cheerless marble-clad place with dark and heavy woodwork, and soon met Tatiana who was to be my interpreter. I had given myself a crash-course in Russian and although it was sufficient for making pleasantries or ordering a beer it was woefully inadequate for anything more.

The next day Tatiana and I met in the hotel lobby and picked our way through the snow to the Ministry of Geology on Lenin Avenue. The officials, ethnic Kazakhs not Russians, were a pleasant bunch: Urbisinov; Akchulakov with his bad teeth but a beaming grin; and Kuantaev his deputy, a younger man with severely disabled hands. I introduced Croft to them, and on handing over a copy of our 1993 Annual Report was immediately congratulated that we were the first company to have presented them with one so up-to-date. But taking a look at the balance sheet, Akchulakov spotted immediately how small Croft was and, laughing, suggested: "Maybe it will be best if we don't talk about the Caspian Sea blocks next to the big companies, yes?"

Within a couple of days we had agreed that the Turgay Basin in the centre of the country was the most appropriate area for Croft. The first blocks they suggested seemed from my maps to include the Baikonour Cosmodrome. Although the Soviet map-makers were notorious for plotting strategically important features in the wrong place – to confuse would-be agressors – I felt it was an area best avoided, and my hosts had to agree that it could indeed present us with operational difficulties. And so we settled on three adjoining blocks further east, and Croft was invited to submit its formal application for a licence. In some respects the geology of the Turgay Basin resembles the North Sea, with a series of fault-controlled basins of Jurassic sediments underlying a folded blanket of younger rocks. A number of oil fields had already been discovered, including the giant Kumkol Field, and such seismic data as I was able to inspect suggested there was scope for finding more.

I made more visits to Almaty through the summer months. On days off it was pleasant to take a taxi into the mountains and walk in the fir forests, sometimes with Tatiana and her young son, and at other times I'd explore Almaty's parks, its wooden cathedral or its funfair and its cramped zoo. A night at the opera was another diversion where a ticket could be bought for a pound or two, although the quality of the performances was apparently suffering now that top musicians rarely visited from Moscow. One of our negotiating sessions was due to be held at the Sanitarium Alatau, on the western outskirts of Almaty. I was surprised that the government officials should suggest a hospital in which to meet, but when we arrived there we discovered it to be a vast accomodation complex with echoing corridors, unbroken views of the mountains and a range of run-down sporting and entertainment facilities. We learned that in Soviet times, deserving party-members would be awarded a period of rest and recreation in a sanitarium – a doubtful pleasure which reminded us of the old joke: "First Prize, a week in Skegness; Second Prize, two weeks in Skegness."

In September the government intimated that it would award us an exclusive licence over our three favoured blocks, and in January 1995 the award of Licence 219 was confirmed. Although it was something to celebrate, it turned out to be the beginning of even harder negotiations. Over the following months, which became years, we discussed with government officials such matters as the split of production in the event we make a discovery; the size of bonuses we would pay when production reaches certain agreed levels; the rate of tax we would have to pay; the amount of oil we would be permitted to export; and the price we would be paid for oil that was sold to the domestic market. The government's expectations were unrealistic and our patience was strained further when they proposed that one of our negotiating sessions should be in Istanbul – at which at least a dozen of their officials would need to take part, they said. They seemed to view Istanbul as Australians might view London, or francophone nations Paris, perhaps reflecting their ancient Turkic roots. We declined, and our negotiations suffered a further setback from ever reaching a satisfactory conclusion.

Throughout these negotiations in Almaty the centre of gravity of Croft's activities remained the North Sea. The Claymore Field was in full production; we had a small stake in the producing Donan Field operated by BP, and in 1995 the Blenheim Field came into production and added to our income. But the world oil price was still well below

$20 per barrel, and year by year Croft continued to make a loss. Certain of the shareholders made loans to the company, including Harrisons who had already become more deeply drawn into funding Croft than they had intended when they made their initial investment. Costly appraisal wells were being drilled on the Kyle oil discovery and we had already farmed out part of our interest to get our share of those costs carried. We urgently needed either to find a company with which to merge or find an investor willing to inject substantial capital.

Hardly a week went by that I did not sit with one company boss or another and discuss a possible merger. The list is long, but one meeting in particular sticks in my memory. It took place in the summer of 1991 and the object of our interest was a small Irish company called Atlantic Resources. The attraction to Croft was that they were quoted on the Unlisted Securities Market (the USM), while they in turn were attracted by our growing portfolio of North Sea and international interests. The logic was that if we merged the two companies it should be possible to raise substantial new capital from our combined shareholder list to fund our further growth. After preliminary discussions in London with their Managing Director, Iain Harrison and I took a plane to Ireland to meet their Chairman. Tony O'Reilly was already a well-known figure. He had played rugby for Ireland before becoming the head of Heinz foods, then acquiring Waterford Wedgewood and the Independent group of newspapers.

We drove up to the front gates of Castlemartin, some twenty miles southwest of Dublin, pressed the intercom button and announced ourselves to an invisible staff member. The gates swung open automatically, and we approached the Georgian mansion along a tree-lined drive. At the door we were greeted by a liveried butler who ushered us into a drawing room where Jim and Jerry were waiting for us. We learned that they were old friends of O'Reilly's and held key positions in his business empire; we wondered if they had been invited along to provide an independent appraisal of us – or maybe they had just had a meeting with the boss and were on their way out; we were not sure. After an hour or so the Chairman entered the room and settled onto a sofa. He was dressed in a shamrock-green jacket, white trousers, Gucci loafers, and was sockless.

For the next couple of hours, in a relaxed and enjoyable exchange of views, we talked about our respective companies, our business strategies, the depressed price of oil, the recent US-led attack on Iraq, and the state of the world. It was then time for lunch, and he led us into a small dining room. Whereas the parts of the house we'd seen so far were sumptuously furnished and decorated, the room we were now in was decorated in the style of a Little Chef roadside diner. An attractive young woman rose to be introduced to us and we learned she was Chris Goulandris, O'Reilly's fiancée. She was a sweet girl and only later did Iain whisper to me that she was the heiress to a Greek shipping fortune. The conversation over lunch was lively and relaxed. Tony O'Reilly regaled us with amusing stories from his days in the Irish rugby team, and in other reminiscences we soon realised that when he mentioned 'Margaret this' or 'Henry that' he was referring to his association with Margaret Thatcher or Henry Kissinger.

He was as charismatic and larger-than-life as his reputation had led us to expect. But our day was not yet over. Before he would let us go he wanted to show us the highlight of his estate. We climbed into his Bentley and he took us to the Fifteenth Century chapel that he had recently restored and in which his father and mother had been buried. By this act of piety, we inferred, Tony hoped to dampen any objections the Church may have to his planned divorce and his forthcoming marriage to Chris. It was a memorable day, but like all of our other merger discussions we were unable to agree on relative values and it came to naught.

While the search continued for a solution to our funding problems I made periodic visits to Guildford to attend operating-committee meetings with our partner Arco, operator of the Blenheim Field. Unusually in the North Sea, the field was being produced via a floating facility rather than the more normal fixed production platform. The reason, of course, was that the field was small and its reserves could not justify the cost of a platform, whereas a "floating production, storage and offtake" vessel (an FPSO) could be used elsewhere once the Blenheim field had been depleted. The man in Arco who headed the commercial negotiations behind this scheme was a young Canadian named Mark Beacom.

One day he called me. He was convinced that a similar scheme, using an FPSO, could be used to exploit many North Sea oil discoveries which, because of their small size, were at present undeveloped. It was an interesting notion and we agreed to hold a second discussion. I invited him to the Oriental Club where I was a member, and he convinced me that there was something in what he said. He was looking for an oil company which would back him.

I took the idea to Croft's board and in spite of our difficult financial position we agreed to recruit Mark to our team. In 1994 we set up a new subsidiary, Croft Offshore Oil Limited (shortened to COOL), and Mark was appointed its Managing Director. From a small office off Trafalgar Square he got to work. He negotiated with a shipping company to convert one of its tankers to an FPSO which COOL would then lease; he spoke with a number of other companies working in the North Sea and concluded a joint-venture agreement with the Brazilian state oil company, Petrobras; and he negotiated with BP to buy their share of certain of their undeveloped North Sea oil discoveries. I was Chairman of COOL and although I was involved in all of these matters it was Mark himself who took the initiatives and negotiated the various deals. We called the scheme "Project Dolphin."

Having failed to find a company with which to merge on terms acceptable to Croft, our situation had become increasingly difficult. We needed outside help to identify investors and we believed that success was more likely to be achieved in the USA than anywhere else. We had acquired a lot of experience of UK financial institutions, and over the years I had made presentations to many of them in London and Edinburgh, with mixed results, but within Croft we had little experience of their US counterparts. In the spring of 1996 I was introduced to a person who it was thought would be able to help us. His name was Michael Nosworthy, a lanky and somewhat goofy-looking old-Etonian. He was a likeable man and over several meetings he convinced me and Croft's board that he had the contacts and the experience to be able to assist us. In June of that year I signed an agreement with his company, Proteus International. Under the agreement Proteus would receive a monthly retainer and a success fee in the event he secured financing for Croft and its activities. With Michael's help we drafted an "Information Memorandum" which was to form the basis of Croft's approach to possible investors. Armed with copies of the Memorandum the three of us, Michael, Iain Patrick and I, flew to the USA. Michael had arranged a series of meetings with institutions he had dealt with in the past. We zigzagged across the continent, making presentations to fund managers in New York, Boston and Houston, but the meeting I remember most clearly was the one in Los Angeles. It was to be a lunch meeting and we flew across the continent especially for it. Near the top of one of LA's skyscrapers we were ushered into the oak-panelled dining room where the decision-makers of Trust Company of the West took their luncheon. A large picture window looked northwest and seemed to be filled by the often-photograhed hillside sign: HOLLYWOOD. It was California at its most glitzy, and the managers were the most unctuous we had met so far on the trip.

We were asking investors to fund Croft to the tune of $US25 million. That would enable us to consummate the purchase of the oil interests from BP and commence the Dolphin Project, and to purchase larger interests in the Claymore Field and the Kyle oil discovery. (As we were already partners in Claymore and Kyle we had a right of pre-emption over any disposal the other partners may wish to make). Although our US presentations did result in some of the institutions seeking further information, and telexes and faxes flashed between Glasgow and the USA, only one came back to us with a firm proposal. It was the Trust Company of the West. We started negotiations with them but at about the same time, November 1996, I met a London-based Canadian called Peter Braaten who expressed interest in Croft's portfolio of interests and in the opportunities open to it. Over the following days and weeks we negotiated a possible tie-up between Croft and his company, MMRL, while ensuring that we kept open the possibility of a deal with Trust Company of the West if acceptable terms could be agreed.

By February 1997 our negotiations with MMRL had reached a point which satisfied the Croft board. Throughout, I had kept Michael Nosworthy informed of the progress of our negotiations and on the evening that Croft accepted MMRL's terms I called him. "Well, I saw it coming" he said, a note of disappointment in his voice that we had now discontinued our discussions with Trust Company of the West. The next day I received a fax from Michael in which he calculated that his company, Proteus, was due a success fee of over US$1.2 million, plus warrants. The Croft board was dumbfounded.

The agreed deal with MMRL was that they, partnered by another Canadian oil company, Bow Valley, would acquire Croft for 3.5 pence per share and would acquire the outstanding loanstock at par. In the opinion of Croft's board Proteus was not entitled to a success fee at all, as they had failed to find a party to "finance" Croft. At least, they had failed to find a party which would finance us on terms acceptable to our board – the terms being offered by Trust Company of the West being completely unacceptable.

I contacted Peter Braaten and told him about our dispute with Michael. He was as outraged as we were. Although we obtained legal opinion that Michael's claim was groundless, there was nevertheless now a contingent liability on Croft to settle the claim. Peter considered this, and a modified offer was put forward by MMRL and Bow Valley. They would pay 3.0 pence per share and would place the remaining 0.5 pence per share in an escrow account pending resolution of the dispute with Proteus.

And that is almost the end of the Croft story as far as my involvement was concerneed. The sale of Croft went ahead, and before the summer was over I was unemployed. The dispute with Proteus dragged on unresolved, although it was now the new owners of Croft who had to reach a settlement. As a former shareholder, hoping one day to receive my outstanding half-penny per share, I contacted Michael from time to time to see if any progress was being made, and at one stage it looked as if a settlement was imminent.

But then, a few weeks later, I received an unexpected call from Michael. His voice was unusually subdued, and tense with emotion: "Andrew has just been murdered."

I could scarcely believe what I was hearing. Andrew was an American who lived in Houston and was Michael's partner in Proteus. In fact it was with Andrew that I had signed the original Croft/Proteus agreement. Michael explained that Andrew had been advertising his car for sale; he was at his Houston home when two men knocked at his front door and asked to look at the car as they may wish to buy it; Andrew withdrew to pick up the car keys and as he returned to the front door the two men shot him at point-blank range, took the keys and drove away in his car. I was shocked and aghast at the news and offered Michael such words of sympathy as I was able to muster.

Andrew's murder meant that his estate and his entitlement to his share of any eventual settlement with Croft were now in the hands of his executors. Apparently they were unwilling to settle for anything less than the full amount of Proteus's claim. Years passed before I was informed that a settlement had eventually been reached. But by then the legal costs had mounted to the extent that only a small fraction of our half-penny per share remained for distribution to us former shareholders.

Croft had given me the most exhilarating thirteen years of my life. Enormous fun and satisfaction had been interspersed with moments of disappointment; there had been times when I felt I was walking on air but others when the sheer hard work and strain had almost made me want to give up the whole thing. Our timing could hardly have been worse, the world oil price plunging from nearly $30 to just above $10 within weeks of us setting up the company; only for a short period, at the time of the first Gulf War in 1991, did it spike above $20. This affected the appetite of institutions to invest in a small oil company and, of course, diminished our revenue once we started producing oil. And the value of Croft when we finally sold it was based on these depressed oil prices. Some shareholders made a modest return on their investments, sadly others made a loss, and I emerged with about the same as I had put in. Whether I could have run the company more effectively over those years is a question I often ask myself. If it failed to grow in the way I'd hoped, perhaps it was at least partly my fault. Maybe I was prone to become fascinated by the geology of projects and lost sight of their commercial realities. And maybe I mishandled the purchase of the substantial producing asset which could have underpinned the flotation of Croft on the stock market – for example, we narrowly missed buying a stake in the Victor Gasfield from Mobil. But of one thing I'm certain:

the team I worked with had been thoroughly loyal, competent, hard-working and supportive to the end. I would not have missed the Croft experience for anything.

Chapter Eighteen

Now aged 61, I used my accumulated funds to buy an annuity and I became a pensioner, exactly forty years after beginning my career.

With time to indulge my liking for travel I decided to visit Sally. Having obtained her degree at Birmingham and worked and travelled in Britain and the USA, she was now living in Australia. She had left England some years earlier and travelled with a friend through Southeast Asia intending to reach New Zealand, the country of her birth. But on the way she met a young man named Chris in Australia, and in due course they married and settled in Perth. When Sally and Chris married I was not invited, and this was something which hurt me deeply at the time. It was the same when Helen married – I was pointedly excluded. There is no doubt that had the girls not been pressured by their mother, things would have been different, but I was disappointed that neither of them was willing to stand up to her and overrule her. Anne must have thought she had finally triumphed in her campaign to alienate me from my daughters.

It was against that background, feeling that our relationship needed nursing back to health, that I went to stay with Sally. She and Chris and their two sons, Oliver and Nick, had a house in Kalamunda, an attractive small town in the hills east of the city. They made me as welcome as I expected they would; we barbecued, drove to see the sights in the Darling Ranges, swam at the local pool and walked the dogs in the eucalyptus and banksia bush which spread from the end of Valley Road. It was a happy visit, but I left feeling more sorrow than resentment toward Anne, sorrow and puzzlement that an intelligent woman could devote her life to a battle where her own children were the main losers.

On the way to Perth I'd seen Jair in Bangkok, and moving on from Perth I stayed with Richard and Sujatha at their home in the Dandenong Hills, the cool and leafy place I remembered so well where the rosella parrots still flew down from the gum trees to be fed on the verandah. I continued my re-acquaintance with Australia on the Gold Coast where Rod and Rudi were now managing an apartment block, Rod as keen as ever on his daily swim and game of tennis. As I have explained earlier, I'd grown very fond of Australia when I lived there in the 'seventies. Returning after some twenty years, I had the unusual experience of feeling the country was even better than I remembered. There was a liveliness to the place, a buzz, and the people one met had a confidence and pride which were new and attractive.

It was late 1997 and back in Britain I found that former colleague Charles Westwood, now a consultant, wished to talk with me. MMRL and Bow Valley had not been the only Canadian companies keen to obtain a foothold in the North Sea through Croft. One of the possible merger partners I'd met a year or so earlier had been Rigel Energy, and Charles let me know that they were seeking someone to join their board and become Chairman of their UK subsidiary. It was an attractive prospect. I met Rigel's Chairman and Chief Executive in London and shortly after was told that they would like me to take up these non-executive positions. The executives in Calgary and London were a pleasure to work alongside, and two of them became lasting friends, John Hodgins and Mark Smith. The quarterly board meetings in Calgary gave me the chance to learn about the Canadian oil and gas business, and provided opportunities to travel in that part of the continent. But Rigel's own cash flow was under stress because of the low oil price, at times touching $10 per barrel, and in 1999 they succumbed to a takeover bid from another, larger, Canadian oil company, Talisman. Again, I was out of a job, and this time it was for good.

I'd thought for many years that if, or rather when, I eventually found myself no longer working I would miss the oil industry. But I surprised myself by not missing it at all. I was only mildly interested in what the oil price was, and days would go by without my bothering even to check it in the newspapers. And news of which company was acquiring which barely made me look up. Much more important were the good friends I'd made over those years, and I still had them. And my love of geology remained as strong as ever – and now I had all the time in the world to indulge it, as well as my other passions.

A holiday with family and friends on the canals of Burgundy in 1992, and in the Dordogne three years later, had sowed in John and me a delight in the countryside of southern France, and it prompted a series of walking trips over the following years. Usually we followed a "GR", a *sentier de grande randonée,* one of the network of well-marked and documented long-distance footpaths that criss-cross France. Our first, in the spring of 1998, was in Provence, starting at Aix and including a trek along the rubbly summit ridge of Monte Sainte Victoire, made famous by Cézanne, and finishing (admittedly with some intervening road transport) at the attractive fishing port of Cassis. It was a trek marred by a crippling bout of dizzineness which overcame me on the second day, scaring the wits out of John and giving me a brief experience of France's excellent hospital service. Whether it was brought on by the fairly severe dehydration we suffered (through our own fault) the previous day on Monte Sainte Victoire or by some other cause, we didn't discover. Surprisingly it didn't put John off further treks with me and the next year, again in the spring, we set off from the western end of the Pyrenees and walked – much of the time in mists swirling in off the Atlantic – till we reached St-Jean-Pied-de-Port, the mountain crossing point for pilgrims heading for Santiago de Compostella and hence liberally adorned with the sign of the scallop shell.

Jair's four-year tour of duty in London came to end in 1997 and she was posted back to the Ministry of Foreign Affairs in Bangkok. Leaving aside my sadness over her departure, a practical implication was that I would have nowhere to stay on my frequent trips to London. Foreseeing this, I scoured the SW London districts for a small *pied à terre* and settled on a third-floor flat being built beside the river in Putney. I watched it gradually take shape, and took possession in March of that year. After she had left, Jair and I continued to meet whenever the opportunity arose – in New York while she was attending a United Nations conference; in Istanbul; and once in Vancouver, as well as on my occasional trips to Thailand. However, after being based in Bangkok for two years Jair returned to the UK in 1999, this time to undertake a post-graduate course in Human Rights Studies at the University of Essex. But by then my life had moved on. We remained on friendly terms, at least initially, but she was a true Thai patriot and career woman who saw no future for herself outside of Thailand.

During that period of gradual estrangement I met Nok. To be exact, I met Nok again, as I had met her several times over the years. She was the sister of our good friend Sri, the Thai wife of my sometime sailing companion, Gordon. Sri had been keen to set up a business in Scotland and she had now decided to open a gift shop. Nok was to be her representative in Bangkok, buying and expediting goods from there, but late in 1997 she was in Scotland, applying her strong sense of style to help design the shop and set out its stock. Along with other friends and dignitaries, I was invited to the official opening of the new venture.

It was a strange, erratic relationship. I was still seeing Jair whenever I could and I sensed that when Nok was back in Thailand I was not uppermost in her mind. But we enjoyed each other's company and made some interesting trips together. I invited Nok to join me on a trip to southwest France the following summer. John and I had agreed we would look for an old farm building there, to buy and restore. The four of us – John at the time in a liaison with Jill – drove hundreds of miles in an ill-conceived and hectic search across half a dozen *départements*. We bought nothing on that trip, but the urge overcame me again three years later – but that's another story.

Later that year I decided I would make good a large gap in my experience and visit East Africa. I mentioned it on the telephone to Nok in Bangkok and she asked if she might accompany me. We rendezvoused in Nairobi in December, hired a Mitsubishi 4WD, a tent and other camping gear and set off northwest. It was a trip full of interest and adventure and Nok was impressively tolerant of the discomforts and unfazed by our brushes with danger. Early in our expedition, in the middle of nowhere our vehicle got bogged down and it was hours before we found a way of continuing; and we never knew in advance where we would spend the night, even high in the Aberdare Mountains with mists swirling and darkness falling. And we had some alarming encounters with locals keen to rob or cheat us. An interesting example of the latter happened in Nairobi soon after I got there and was waiting for Nok to arrive from Bangkok.

I was downtown, wandering the crowded streets and taking in the sights and the local colour, when a young African beside me asked if I was on holiday in Nairobi. It seemed a friendly enough question and we got into a conversation.

"Look" he said, "I wonder if you can help me. I'm keen to go to England to continue my studies, and I'd really like to know more about the facilities there, and the education system."

"Well, I'm not an expert, but I'll try and help you." I said, "What do you want to know?"

We continued like this for a while, until he suggested we should go and have a drink together at a nearby café where we could talk more comfortably. I agreed, and followed him, threading through the crowds until we came to a nondescript, open-fronted café on the next block. There were several empty tables and I motioned him to sit down while I moved toward the counter to buy a couple of orange juices. But when I returned he was still standing, and said "No, don't let's sit here, it is much better upstairs." I was puzzled by this and demurred, asking why we should go upstairs when there were plenty of empty tables below, where, additionally, we could enjoy watching the passing crowds.

He reluctantly accepted and we sat down. As his story unfolded it became clear that he was not planning to study in England at all, but was seeking help to travel from Nairobi to his home in one of the townships in South Africa. I continued listening for a while, for he seemed a pleasant young man and was unthreatening. But he persisted, albeit scaling down his request from two-hundred pounds to fifty. Finally I grew tired of his entreaties and told him that I'd been happy to talk with him, but I have a rule never to part with money to people I've not met before. I left him sitting in the café and strolled away, thinking little more about it.

Fifteen minutes later, while I was standing at a busy crossroads waiting for the traffic to thin so that I could cross, I felt a tap on the shoulder. This was another man, African, in his thirties and well-dressed in a dark suit.

"Were you just speaking with a black man – in a café?" he asked.

"Yes, I did have an orange juice with a young chap, and we sat and talked for maybe twenty minutes." I replied, puzzling over why it should be of interest to this newcomer.

"Anyway, who are you?" I asked, "and why are you questioning me like this?"

He didn't answer immediately, but continued with his interrogation:

"Did you give him any money?" he went on.

"No, I certainly didn't, although he did ask me for money." At this point he identified himself:

"I'm with the Kenyan Security Service" he said, and produced an ID card from his pocket and showed me. Then another dark-suited man appeared, and was introduced to me briefly as his colleague. They conferred, before saying:

"This is a serious matter. The man you were talking with is now in our custody and when we picked him up he was carrying a very large sum of money on him. I think, if you don't mind, we'd better find somewhere to sit quietly where we can talk this over."

Beginning to think what an unusual city I'd landed in, I accompanied the two men. We walked for a couple of blocks and then entered a quiet restaurant. They guided me to a table in the corner, out of earshot of anyone. "What are you doing in Nairobi?" they asked, and "Where are you staying?" The interrogation continued in this vein, and I was asked to take out my wallet and explain where I had obtained the sterling banknotes I was carrying; apparently my erstwhile acquaintance was found to be carrying US dollars. They weren't hostile or discourteous, but they were persistent. After ten or fifteen minutes they seemed satisfied. They apologised for the inconvenience they'd caused me, explaining that since the bombing of the US Embassy just four months before, they needed to be very careful about who entered the country and what they were upto. I walked off, unsettled and thinking I wouldn't wish to be in that young South African's shoes.

Although we had some incidents in Kenya which, on looking back, could have become ugly, we were also rewarded with some unforgettable experiences. Camping beside Lake Naivasha we woke in the middle of the night to a sound between a lion's roar and a horse's neigh, a strange and chilling noise which the next day we deduced was probably a hippo which was doing the rounds of his landward territory. At Lake Baringo the flamingos formed a stunning pink carpet spreading for a mile across the hot, shallow waters. In the dry Samburu country a zone of green forest fringes the beautiful Ewaso Ngiro River; the game there is plentiful and we saw at close quarters buffalo, giraffe, zebra, many kinds of antelope, large herds of elephant – and a crocodile emerged from the water at dusk to watch us as we dined on the terrace of the Buffalo Springs Lodge. We camped from time to time, but there was no hint of roughing it for the several nights we stayed at Tree Tops, the game-watching hotel made famous when Princess Elizabeth was staying there and learned that her father, the King, had died and she was now Queen. After Christmas in Nairobi, Nok returned to Bangkok and on Boxing Day I was back in London.

But a tougher and more challenging expedition was in store. In 1997, when I found myself with time on my hands, I decided I'd get in touch with the The British Schools Exploring Society, the society I'd joined in 1954 to take part in their expedition to Northern Quebec. I'd been out of touch with them for nearly half a century but found them housed in an attic above the Royal Geographical Society in Kensington Gore. I rejoined and started receiving their newsletters. Then one day I read that they were inviting members with suitable experience to come forward with proposals for new expeditions. This was not what I had intended when rejoining, but now I started to imagine myself leading a BSES expedition. Within the last few years, while on business trips, I'd been into the foothills of the Tien Shan on the Kazakhstan border, and the idea took root in my mind that this mountain range would be a fine place to lead an expedition. The more I researched the idea, the more attractive it became. But instead of going to Kazakhstan, I would propose an expedition to the highest and most remote part of the Tien Shan, in Kyrghyzstan.

A few days later, over lunch at the RGS, I was explaining my proposal to Jon Fleming, a friendly ex-military man and formerly a Queen's Messenger, but now the society's Executive Director. He was attracted by the idea and agreed to put it before the BSES council. It received their blessing and I started in earnest to plan the 1999 Tien Shan Expedition. In the summer of 1998 I carried out a two-week reconnaissance with Sarah White, the young and capable Expeditions Officer who worked alongside Jon. With the help of Vladimir Birukov and his company, Tien Shan Travel, we satisfied ourselves that the expedition would be feasible. Finding a basecamp site which combined a large flat area for tents, with access to clear as opposed to glacial water, was surprisingly difficult but we were eventually successful.

Back home in Scotland, the most difficult task I faced was assembling a team of suitable leaders with the range of scientific and mountaineering skills we would need. But I was fortunate in finding fourteen capable and enthusiastic men and women willing to give up their time and join me the following summer.

This was the society's first foray into Central Asia, and the lure was of course the magnificent snow- and ice-covered mountain chain which runs the length of Kyrghyzstan, and makes up the Tien Shan. It bounds the north of the Taklamakan Desert of China and is universally known by its Chinese name, meaning the Celestial Mountains. At the eastern end of the chain are peaks upto 7000 metres high, and it was into this region that I decided to lead the expedition. It was mid-July 1999 and we set up basecamp on a flowery meadow where the clear waters of the Tyook side-stream join the mighty, putty-coloured Sary Djaz River. We got there from the town of Karakol, 130 kilometres away, by six-wheel-drive Kamaz trucks, ex-Red Army vehicles tough enough to cope with the pot-holed and partly washed-out track which snakes its way through the Tien Shan, reaching a height of 3822 metres where it crosses the watershed at the Chok Ashoo Pass.

The upper Sary Djaz valley is close to the Chinese border and so it is a militarily sensitive area. I discovered this at 2.30 a.m. on the first night, a night of hard frost and a million stars, when I was awakened by noises a few metres from my tent. Getting out to investigate I was dazzled by a bright light held in my face, but I could see I was looking up the muzzle of a Kalashnikov, and there were men in camouflage uniforms around me. It was a horse-mounted night patrol from the military base at Echkillitash, further up the valley. With the help of Farida, our interpreter, it took thirty minutes to subdue their hostility and to persuade them that we were not an invading army – an easy mistake for them to have made considering we had some two dozen tents spread across the valley floor. If I had not been able to produce a letter from the President of the Republic, the Patron of our expedition, I suspect it would have taken rather longer to persuade them.

With our credentials established, the expedition settled in, using the first few days to sort out gear and get acclimatised to the 3000 metre elevation of basecamp. It was then time for the real business to begin. Our 57 young explorers drawn from schools across the UK and with some from Kyrghyzstan, were split into smaller groups, each of them under a pair of leaders with expertise in different field-sciences: geology, glaciology, topographic survey and botany. While some of the groups went about their fieldwork, the others trekked south to gain mountaineering experience up on the glaciers at what we called Mountain Camp. Getting from basecamp to Mountain Camp with all of the gear and food to last a week was a two-day grind over bouldery river beds, boggy steppe and finally up the face of a formidable moraine, but once there, the magic of the place worked its spell on everyone. Beneath the snow-covered Sary Djaz peaks our cluster of tents at 3800 metres was dwarfed by this vast barren amphitheatre of glaciers and moraines.

By the second week of August much of the fieldwork was complete, everyone had learned the techniques of travel over snow and ice, and we were all a good deal fitter – although even at basecamp I found I panted at the least exertion. It was time for the third phase of the expedition, the Long Treks. These had been a tradition of the Society but had been dropped from recent expeditions, I suppose because they smacked of elitism, but I thought that was bunkum and so decided to re-introduce them. We settled on six routes, of different degrees of difficulty, and the youngsters were allocated according to their fitness and, more important, their enthusiasm. I led the easiest of them, a five-day trek, at times through blizzards, north to the watershed of the Tien Shan where we could stand with one foot on the slope toward China and the other on the slope leading north to Russia. The other Long Treks were much more challenging, a couple of them lasting nearly a fortnight and spending long periods high on the South Inylchek Glacier.

It was while returning down the glacier some nine days after they had left base-camp that I got news by HF radio that Anna, one of the leaders, was seriously ill and needed emergency evacuation. Often the radio reception was too bad to make contact but on this day, as luck would have it, reception was good. I managed to radio the news to a mountaineering camp on the other side of the range and within the hour their helicopter was picking up Anna from her camp on the glacier and taking her to hospital in Karakol. An emergency appendectomy was carried out there and Anna was afterwards taken by air ambulance to the UK where she made a full recovery. It is a sobering thought that had the HF radio not worked that day we would probably have had a tragedy on our hands.

With all of the treks returned, it was time for the expedition to pack up and make its way back to Karakol and from there to Bishkek. In the capital we were privileged to be invited to a reception at Government House by the President of Kyrghyzstan, Askar Akaev. Not put off by the malodorous scruffs assembled before him, the President, himself a research scientist before he entered politics, spoke generously about the British and the work done by the Society. I responded with a speech, equally fulsome in its praise of Kyrghyzstan and its people, and we then exchanged gifts before starting on the sandwiches and chocolate cakes. I still have the volume of Manas's epic poems he gave me, but I was sad to see that he was ousted in an uprising a few years later and was forced to take refuge in Moscow.

From my own BSES expedition to Northern Canada nearly fifty years earlier, I knew the beneficial effect that a tough spell in the wilderness can have on a youngster, encouraging resourcefulness, determination and self-awareness. And now I'd seen its transforming effect from a new perspective. The Tien Shan expedition led to my becoming closely involved with BSES and its charitable work, and I was elected onto the BSES Council and became chairman of its Expeditions Committee.

Before the end of the Millenium I squeezed in a second walking trip in the Pyrennees, this time with my two Canadian friends from Rigel, both of them – like myself – made redundant by the takeover of the company. Hitherto, my acquaintance with them had been in a business setting, but they turned out to be good-natured and amusing trekking companions. The plan was to set out from Vernet les Bains in Rousillon and end on the Medittetanean coast at Banyuls. The route took us high over the flank of Pic du Canigou, at 2784 metres the highest point in the eastern Pyrennees. It was late autumn and Canigou was snow-capped and icy. After a couple of days we were making for what the guide-book led us to believe was a hostel on the mountain. As darkness fell we found ourselves at the snowline, the ground frozen solid, and the hostel coming into view. We stumbled in expecting to be offered hot drinks and a comfortable bed, but were disappointed – nay, aghast – to find that it was no more than a mountain hut. We had no choice but to stay, even though we were not equipped for the coming night. There was no water in the hut and we had drunk the last of the water we had been carrying, and so we were reduced to scooping snow from the few millimetres that covered the ground and melting it as best we could. There was a fire-place and we did manage to start a fire, but soon ran out of logs and so retired upstairs. There we found wooden platforms on which walkers would normally have spread their sleeping mats, but we had none. It was a horrible night, and the periodic sound of shuffling and grunting showed the others were getting as little sleep as I. At first light I looked around the icy dormitory. I could see Mark was awake, and as he saw me he lifted his head and looked at me. In a voice that was full of suffering he moaned: "Oh God, I hope they take us to the gas chambers today!" Politically incorrect, certainly, but whenever I think of it I can't help chuckling.

The new Millennium was approaching with much hoo-hah. We had been led to believe that as midnight rung out on 31st December the "Millenium Bug" would strike, the world's computers would crash and chaos would ensue. Companies and governments paid large sums to consultants to minimise the risk, and I recall that cost-updates and progress reports had been regular agenda items at Rigel's board meetings. There was to be a huge fireworks display on the Thames, some kind of extravaganza at the ghastly Dome at Greenwich, and people started planning their celebration parties.

But I'd made up my mind how I would see in the New Millenium, and it would be a Hogmanay I'd remember. On the last day of 1999 I drove from Millfaid, over Rannoch Moor and through Glencoe to the West Coast, and parked beside the boathouse where we always parked when visiting our tiny cottage by the shore of Loch Linnhe. I was not alone. Erskine the family dog was all the company I needed. Snow lay on the ground on the drive north, but now it was raining hard as we walked the mile along the coast in the gathering darkness, stumbling through bogs and over heathery hillsides till we reached the door of Leachnasgeir. Inside I found dry wood and got the fire alight, then cooked a simple spaghetti meal. And we sat there, Erskine and I, our wet coats steaming before the dancing flames, listening to BBC Radio 4. At nine o'clock I unrolled my sleeping bag and turned in for the night, Erskine faithfully curled up on the floor beside me.

The Millenium Bug failed to materialize and the world's doomsters switched their apocalyptic predictions to global warming and the dogma that it is caused by mankind. Sticking my neck out, I predict that the "greenhouse effect" will prove to be a theory which, while elegant and seductive, is of much less importance than the natural variations in climate that have taken place over geological time.

In mid-January I returned to London to address several hundred youngsters and their parents on the 1999 Tien Shan Expedition. It was held, as it always has been, at the Royal Geographical Society and Prince Andrew, the Patron in Chief of the British Schools Exploring Society, presented me and each of the other Chief Leaders of 1999 expeditions with a commemorative shield. Mine now hangs on the bathroom wall in the London flat.

With trips planned to go down the Mekong River, and another assault on the Pyrenees with John, the year 2000 was shaping up as not much different from the others. But then in June I met Mikiko.

Printed in the United Kingdom
by Lightning Source UK Ltd.
121385UK00002B/56/A

9 781905 529483